FOREST PLAN

ECOLOGICAL INTEGRITY *and the* MANAGEMENT *of* ECOSYSTEMS

Edited by
STEPHEN WOODLEY
Canadian Parks Service

JAMES KAY
Environment & Resource Studies,
University of Waterloo

GEORGE FRANCIS
Environment & Resource Studies,
University of Waterloo

Sponsored by Heritage Resources Centre, University of Waterloo,
and Canadian Parks Service, Ottawa

Manufactured in the U.S.A.

Library of Congress Cataloging-in-Publication Data
Ecological Integrity and the Management of Ecosystems/edited by Stephen Woodley, James Kay, George Francis.
p. cm.
"Sponsored by Heritage Resource Centre, University of Waterloo [and] Canadian Parks Service, Ottawa."
Includes bibliographic references.
ISBN 0-9634030-1-X (hardcover)
1. Ecosystem management—Congresses. 2. Ecological integrity—Congresses. 3. Ecosystem management—Canada—Congresses. 4. Ecological integrity—Canada—Congresses. I. Woodley, Stephen J. (Stephen Jerome), 1951- . II. Kay, James. III. Francis, George. IV. Heritage Resource Centre. V. Canadian Parks Service.
QH75.A1E24 1993
363.7—dc20
93-6949
CIP

Contents

APPENDIX

Preface

As society's dependence on resources wrought from ecosystems deepens and therefore, the stress under which those ecosystems are put increases, the need to assure ecological sustainability has become widely recognized. This, in turn, has led to efforts to rehabilitate badly degraded ecosystems and protect areas which have important ecological value, such as national parks, critical fish and wildlife habitats, particular natural communities, or components thereof, i.e., endangered species.

Human values are an integral part of the decisions to protect or rehabilitate, but the goals and objectives for such actions are often implicit or unclear. Ecosystems are valued because of certain features they contain or functions they serve. Agencies involved with rehabilitation measures in, for example, the 43 badly degraded "areas of concern" along the nearshore areas of the Great Lakes, contrast ecosystem degradation with ecosystem health and refer to their actions as restoring *ecosystem integrity.* The conservation of natural areas is justified in terms of maintaining ecosystem integrity and biotic diversity. The concepts of "health," "integrity," and "diversity" express important values associated with the management actions being undertaken, but they do not provide clear guidelines for these actions.

As agencies find themselves under legal or policy commitments to assure the existence of ecosystems that express these values, the need for definitive guidelines and criteria is urgent. For example, the 1987 U.S. Canada "Great Lakes Water Quality Agreement" stated that its purpose was "to restore and maintain the chemical, physical and biological integrity of the waters in the Great Lakes Basin Ecosystem." A 1987 amending protocol reconfirms this purpose and calls upon the parties in the agreement to define lake ecosystems for each Great Lake (or portion thereof) and to specify ecosystem health criteria to evaluate progress towards these objectives.

In August, 1988, the Parliament of Canada passed an amendment to the National Parks Act requiring that "maintenance of *ecological integrity,* through the protection of natural resources shall be the first priority when considering park zoning and visitor use in a management plan." Inclusion of the term "ecological integrity" as a legal responsibility has raised the problem of usefully defining the term in the context of National Park management.

Another example of the efforts toward managing, in accordance with more definitive ecosystem-based concepts and criteria, comes from the United States Forest Service. They are legally mandated to protect biological diversity, but there is an ongoing controversy as to how this should be done.

This search for ecologically sustainable management or ecological integrity involves both quantifiable measures of ecosystems as well as social or value judgements of what is being sought. There may be some examples of ecosystems in which there is no argument about integrity. All involved would probably agree that a pristine park in the north Yukon has ecological integrity. It has a full complement of native species, ecological processes and structures and high quality water and air. We may also agree that a corn field in southern Ontario lacks integrity. As an ecosystem it has loss of diversity, ecosystem functions are impaired (e.g., rapid nutrient loss) and structures are rapidly degrading (e.g., soil loss). Similarly, there would probably be agreement that much of Lake Superior has retained the integrity of its aquatic ecosystems, while Toronto Harbour has not. However, beyond the extremes the answers are less clear. Does ecological science have the ability to provide useful measures of ecological integrity? Is there a useful process for incorporating value judgments within measurable ecological variables? This volume addresses these questions.

This effort to confront these issues brings together individuals from a variety of backgrounds: academics, government policy makers, park managers, environmental group members and diverse readers — all with a stake in ecosystem management. The editors hope this text serves as a contribution to the solution of the difficult question of living in harmony with our planet.

Setting the Stage

The Notion of Natural and Cultural Integrity

HENRY A. REGIER
Institute for Environmental Studies, University of Toronto
Toronto, Ontario

Introduction

The notion of ecosystem integrity is rooted in certain ecological concepts combined with certain sets of human values. The relevant normative goal of human-environmental relationships is to seek and maintain the integrity of a combined natural/cultural ecosystem which is an expression of both ecological understanding and an ethic that guides the search for proper relationships.

A living system exhibits integrity if, when subjected to disturbance, it sustains an organizing, self-correcting capability to recover toward an end-state that is normal and "good" for that system. End-states other than the pristine or naturally whole may be taken to be "normal and good."

I first sketch a characterization of ecosystems which reflects different perspectives adopted by ecological experts, and refer to the features which characterize the self-organization of ecosystems. I then note the particular set of human values associated with maintaining or enhancing these features of ecosystems, in part by contrasting them with other sets of cultural values that guide human activities vis-à-vis nature. Cultural integrity, as I define it, is human capability individually and through institutions to complement the integrity of a modified natural ecosystem in an overall context that is inevitably turbulent, socially and ecologically. At the conclusion of this paper, I have sketched, with the help of colleagues, a general image of an ecosystem in a state of integrity.

From the top down an ecosystem is a part of the biosphere; from the bottom up it is the organism interacting with other organisms and non-living features of their shared habitat. Some may even term the entire biosphere an ecosystem. Others may note that an organism such as a human serves as a habitat for a variety of other species, together with some non-living material as in the gut, so that a single human may rate as an ecosystem. It's an elastic concept — not only with respect to scale.

Integrity and Disintegrity of an Ecosystem

From an Aspect of Natural Ecological Processes.

The concepts of open-system thermodynamics are increasingly being used as a starting point in a discussion of self-organizing and self-integrating processes in living systems (e.g., Ulanowicz 1986). If I were more competent in that field, I might take that approach here. Instead I will use a less theoretical, more empirical starting point, i.e., the generalized sketch by Bertalanffy (Davidson 1983) of how living systems effect self-integration:

> "Bertalanffy's model of hierarchical order [has] four related concepts: As life ascends [from a state of low] complexity, there is a *progressive integration,* in which the parts become more dependent on the whole, and *progressive differentiation,* in which the parts become more specialized. In consequence, the [living system] exhibits a wider repertoire of behavior. But this is paid for by *progressive mechanization,* which is the limiting of the parts to a single function, and *progressive centralization,* in which there emerge leading parts ... that dominate the behavior of the system."

Bertalanffy perceived that these processes occurred within the self-organization of biological (i.e., organismal), sociological, and ecological living systems. Because every living system is phenomenologically unique within its specific historical and environmental context, there is much that such an apparently simple set of statements cannot explain. But this is also true for any other universalistic scientific explication of some living system at whatever level of organization. Here Bertalanffy's sketch is taken to be consistent with ecologists' general descriptions of how "ecological succession" within a benign environment proceeds through various sequences.

Holling (1986) has illustrated how a living system responds to internal processes and to external perturbations or stresses that are "normal" to the system. Here "normal" implies two things. Such a stress is "usual" in that it has a long history of recurrence. It is also "desirable" in that the system's long-term self-integrative strategy has internalized a need for such a stress, at least periodically. Within a general regime appropriate to it, such a stress is good for the living system.

Perception of what an ecosystem is has its more holistic phenomenological aspects (species, communities, land forms, climate) and its more reductionistic analytical aspects (energy cascading, material cycling, information organizing). Each of these classes of features — holistic or reductionistic — has a very large number of possible states or elements that may be perceived. A characterization of an ecosystem cannot be complete — at most it can only be very partial. But, if based on canny insight, such a characterization may prove quite useful.

The implicit emphasis in Bertalanffy's characterization is on *information* as related to self-organizational capabilities. Some earlier analytical attempts to characterize ecosystems focused on material cycles. Various components of an ecosystem were analyzed to determine the amount of water, carbon, nitrogen, phosphorus or other "simple substances" in those components. How these materials cycled through an ecosystem was then described in a kind of accounting context, an input-output model or a dynamic simulation.

Material cycling studies were followed by energy cascading studies. The incoming energy from the sun was perceived to flow through different temporary storages before it was radiated back into outer space. Energy content was usually characterized by the amount of heat generated when the material in a component was burned.

The conventional material cycling and energy cascading approaches have provided some useful, if approximate and simplistic, information about ecosystem phenomena. Both approaches miss the most important aspect of self-organizational ecosystem behaviour. Abiotic materials enter into living things and become part of highly intricate structures and processes before eventually reappearing in abiotic form. High-intensity energy of sunlight is taken up to create and animate the highly intricate structures and processes before eventually escaping from the food web as low-grade heat energy. Within the conventional material-cycling and energy-cascading approaches the living parts of ecosystems are treated, in effect, as black boxes. In this way they miss almost entirely the self-organizational features of ecosystems.

So now we are trying to focus analytically on the self-organizational aspects. The holistic natural historians take the fact of self-organization for granted or so self-evident to their senses and observation as not to warrant any quibbling. But analysts must, according to convention, find a non-sensual way of defining, describing, measuring and explaining such phenomena. So they are reaching for generic Bertalanffian self-organizational principles, Prigoginean energy-dissipating phenomena, Mandelbrotian fractals within chaotic contexts, Thomian catastrophe topologies, Hollingian convoluted sequences, etc. At present, such efforts provide intriguing perspectives for further study and some intuitive images of what a healthy environment is all about, but they have not been made practically operational for routine application. Energy cascading, material cycling, information organizing — these analytical approaches, as yet, all have only limited utility in practice, as interesting as they may be in academia.

Practically useful insight and information flows mostly from comparative empirical comprehension that is a pragmatic synthesis of holistic perception and analytical understanding, i.e., between tempered holism and tempered reductionism. The scientific fields of limnology and fisheries (also of agriculture and forestry) have emphasized comparative empiricism. Parks-related studies have a continuing role for holistic perception. Academic ecology has devoted greater attention to analytic and holistic abstractions usually mobilized as band-wagon paradigms, as described by Kuhn (1970) and Feyerabend (1988).

The comparative empirical approach uses, perhaps rather casually, the conventional scientific method that was used in chemistry, physics, physiology, etc.. Phenomena are observed, described with respect to structure and dynamics, classified, compared, contrasted, and explained causally. With respect to aquatic ecosystems, this process was well underway scientifically a century ago. It has not enjoyed as much attention as it deserves, in part because rampant analytic reductionism and band-wagon paradigmatics have been recurring distractions.

With respect to the management of human activities concerning the Great Lakes, most of the information of practical relevance has been generated through comparative empiricism, or what may be termed a "stress response" perspective. This applies to water quality, fish, water fowl, parklands and also to water quantity and sediments. This approach informs the engineers and technical experts who try to control the loading of harmful substances into the lakes (from sewage works, land run-off, dump leachate). It also informs those who set water quality criteria or benchmarks toward their virtual elimination. Further, it informs harvesters of fish and water fowl, as well as users of beaches and parks.

Comparative empiricism is a kind of hybrid between holistic perception (as by an informed naturalist) and analytic understanding (as by an abstract scientist). Some of all three types of emphasis can be found in government service and in academia, perhaps in comparable proportions.

In Figure 1, the vertical ovoid to the left is meant to depict the self-organizational domain of a particular natural ecosystem. Under the influence of external and internal natural stresses and perturbations, and in the absence of any cultural stresses, the system's set of organizational parameters will, with high probability, fall within the ovoid's boundary. After the system has been subjected to one or more sharply disruptive natural stresses it may be found at a relatively low level of self-integration, i.e., towards the bottom of the ovoid. After years in which powerful disintegrative stresses have been absent, but moderate external and internal stresses have acted, self-integration will have led to a set of parameters near the top of the ovoid. The closer the approach to the top of the ovoid, the more vulnerable the system becomes to relatively severe disintegra-

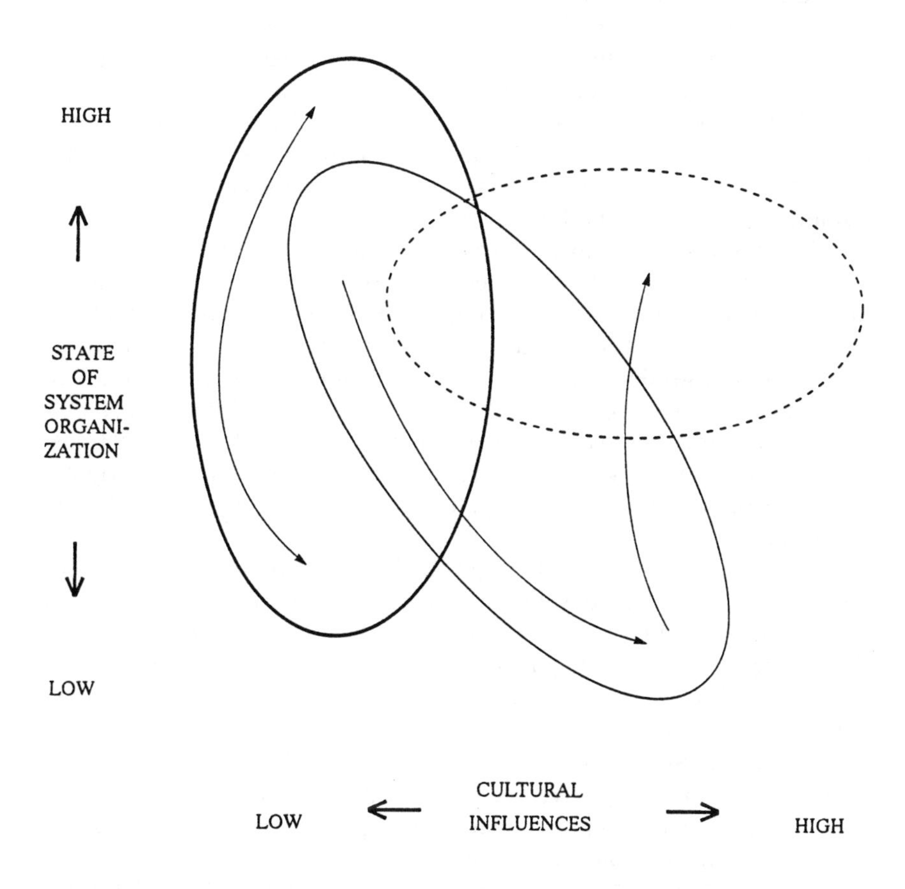

Figure 1. Ovoids refer to three multi-dimensional ecosystemic domains: left vertical ovoid reflects the fully natural state; central diagonal ovoid reflects different levels of ecosystemic degradation due to cultural abuse; right horizontal ovoid reflects a healthy natural/cultural ecosystem or landscape mosaic. The state of healthy self-integration is high at the top of the figure and low at the bottom. The intensity of cultural influences is absent at the far left and intense at the far right.

tion with the recurrence of some natural stress. The convoluted successional/recessional model by Holling (1986) may be seen to be nested within this ovoid.

When human exploitist interests (see below) intervene in such a natural ecosystem, they impose cultural stresses that differ in kind and degree from the existent suite of natural stresses. The intent, after all, is to create something unnatural. It has often been assumed, tacitly or explicitly, that the disintegrative effect of such exotic/cultural stress will be similar to that of some natural stress. Thus, the effect would be to displace the system downwards and either to the right or left, but still remain, with high probability, within the boundary of that ovoid. This is not usually the case, at least if the cultural stresses achieve any measure of intensity. Instead the consequent disintegration has generally been abnormal or pathological (Rapport and Regier 1992).

The ovoid slanting downwards to the right in Figure 1 is meant to illustrate what happens to ecosystem organization with an intense cultural stress. The system is forced far beyond the original bounds and into an unnatural, partially disintegrated state. It may be driven to much lower levels of disintegration than with any natural stress — as when the soil of an urban centre is covered entirely by concrete. Even if the original cultural stress is then relaxed, it may be that the natural system does not possess the recuperative, self-integrative capabilities to recover to the original domain. Such crippling occurs quite generally when exploitist interests have their way.

The horizontal ovoid to the right implies ecodevelopment or sustainable development if the development starts with nature in a pristine state. Or it implies sustainable redevelopment in that human corrective action, directed toward the disintegrative cultural stresses, may lead to system reintegration. But, once severely degraded, a system may have been altered permanently. Some species may have been extinguished and no surrogate — introduced or invasive — can then fully take its place. Or some physical structure may have been altered permanently. Restoration of highly degraded locales to a state resembling the pristine is seldom possible, at least until after the next ice age.

The right horizontal ovoid represents a "healthy," natural/cultural ecosystem. It will continue to be dynamic and will fluctuate between states of higher and lower levels of self-integration, as a function of fluctuating stresses and perturbations. But both high and low states of self-integration will be unlikely because humans will act to prevent both extremes. Desirable structural and process features, towards which such a system may evolve self-integratively, will need to be specified, in part, through a cultural planning and decision process.

With conventional development, the disintegrative effects of the dominant mercantile/industrial/developer interests radiated out from the various metropoles. In the Great Lakes for example, pathological disintegration of the aquatic ecosystems was most severe in the estuaries and bays next to the most industrialized cities. Examples include Buffalo, Chicago, Cleveland, Detroit, Gary, Green Bay, Hamilton, Toronto and Saginaw Bay, — here listed in alphabetic order. Such degraded "areas of concern" (GLWQB 1985, Harris *et al.* 1988) would be plotted toward the bottom of the slanted ovoid in Figure 1 — in the domain of pathological disintegration.

Somewhat above these ecological slums one could plot many other inshore "areas of concern" which are mostly also rivers and bays, such as the Bay of Quinte, Collingwood, Duluth, Thunder Bay and Toledo. Generally less degraded than any of the 42 nearshore "areas of concern" are the open waters of the lakes proper (Regier *et al.* 1988). For all of these debased parts of the Great Lakes Basin ecosystems, efforts are now underway to foster reintegration to more desirable states, which will fall within the horizontal ovoid.

From an Aspect of Cultural Social Processes.

In North America, the 1960s brought with them strong political activism to reform a number of abominations: war, gender discrimination, racial discrimination, class, poverty and environmental abuse. As Sagoff (1988) demonstrates, each of these reforms was based on *a priori*, ethical principles of the general type termed "deontological" by ethicists. Consequentialist ethics, e.g., as provided by utilitarians who had come to terms with "progress," were considered as insufficient at best by each group of reformers. But competition between the various reform groups for media and public attention prevented any effective alliance from emerging. That a common set of ethical principles could be found to which all groups could adhere was a question not addressed. Old-line ideologues, e.g., communists and capitalists, sought opportunistically to co-opt the reformers in the service of their obsolete nostrums.

The mix of values that come to be served by *ad hoc* political decisions within a context of conflict between values is often characterized pejoratively as a "lowest common denominator," LCD. Precisely what is meant by this term is seldom specified, in fact, I have never encountered an explanation of what is meant by it in such a context.

By analogy with the term LCD, as used in arithmetic, it could mean that only those actions which would serve fully all values would be admissible. Such an interpretation would ignore the inevitable reality that some actions serving a particular value would be harmful toward a different value. Thus, an extremely optimistic version of multi-purpose use might be implied by the LCD, and the unrealism of such a hope would lead to the pejorative overtones that come with the use of the term.

Alternately the connotation of the arithmetic term "highest common factor," HCF, may be what is actually intended by the use of LCD. The HCF would be that domain of value in which all the different types of value overlap. Use of HCF as a criterion would benefit each set of users at least in part, but perhaps none fully. The minimalist character of HCF might justify pejorative overtones.

From the context in which the term "lowest common denominator" is used pejoratively, I suspect that the intended meaning, in some cases, is that the most crass or unenlightened value is being served and more sensitive or civilized values are ignored. Neither of the two arithmetic terms, LCD or HCF, would be appropriate for this case, though pejorative overtones would be justified.

In the Great Lakes Basin, there is near unanimous concern about environmental degradation but much less agreement on the kind of values to attach to natural phenomena. For an analytic framework, I have used a classification by Norton (1989) who distinguished four types of values that guide our culture's activities toward nature.

Exploitist interests see nature, in purely natural form, as valueless. Nature does provide raw material, valueless in itself, which may be used by humans to fashion cultural objects which may then be valued. The raw material is perceived as being virtually limitless in quantity or substitutable in quality. This value-orientation to nature is dominant within conventional industrial, market and developer interests. Note that individual humans who serve such interests in a corporate or institutional setting may also be influenced by other, contradictory values as private citizens (Sagoff 1988).

Utilist interests, termed "conservationist" by Norton (1989), see some parts of nature as valuable, other parts as of no concern, and still other parts as harmful. Natural **productive** processes that provide useful resources are valued, as are **assimilative** processes that hide or inactivate our unwanted wastes. Precautions are taken by utilists to safeguard such valued processes and features because they are quantitatively limited. Also, they may seek to control or eradicate natural phenomena which they find harmful. This orientation is currently dominant within interests focussed on: farming, forestry, fishing, water supplies; on waste "management;" and on public health and safety as related to environmental factors. Benthamite guidelines, e.g., "the greatest good to the greatest number — and that for the longest time" may be invoked (Van Hise 1911).

Integrist interests, termed "naturalist preservationist" by Norton (1989), hold that all natural phenomena likely play important and ultimately desirable systemic roles but that not all of our cultural phenomena are "valuable," in the long run. Human culture must ultimately be adaptive to the natural evolving processes, or that culture will be an unpleasant, unhappy and short-lived systemic phenomenon. I have used the term **integrist** to reflect the importance of "land" or ecosystemic **integrity** as urged by Leopold (1949). Integrist interests may favour a landscape, bioregional or ecosystem mosaic guided by large scale zoning. Exploitist interests may be reformed and then accommodated in relatively closed locales of limited size. Utilist interests may be reformed and continue to husband much of the landscape that is ecologically tolerant of utilist practices. Inherentist interests (see below) may be served by reserving parts of the landscape from exploitist and utilist interests in order to remain naturally wild. The integrists may seek a healthy natural/cultural bioregional mosaic in which each of the major legitimate interests plays a fully responsible role.

Inherentist interests, termed "extentionist preservationist" by Norton (1989), ascribe value to living organisms of some (or all) other species (and also perhaps to non-living phenomena) as similar in some way to the value they take to be inherent in a human individual. Natural phenomena have value beyond any use or harm that they may offer to some human interest. People who seek to have parts of the landscape reserved for non-recreational wilderness may be inherentists, as defined here. Also included are people who invoke natural rights to protect some species from human abuse, as they would define "abuse." Religious rituals may be used to atone for a necessary violation of some organism's rights.

In the Great Lakes/St. Lawrence River Basin all of these four types of value orientation are apparent — within society and within individuals. Within the scholarly and scientific disciplines, the neo-classical free-market economists tend to serve rather selfish **exploitist** interests, to the disadvantage of the other interests. Environmental economists and environmental ecologists have found ways to temper exploitist activities so that they do not usually self-destruct within one human generation; but these **utilists** have not found ways to prevent such self-destruction within three generations (Regier and Hamilton 1990). **Integrists,** with prodding from **inherentists,** are trying to limit the scope of short term selfish interests, and the activities of professionals and institutions that serve them. **Inherentists** are exploring deeper cultural phenomena in hopes of discovering ways to achieve more radical reform of our culture — meanwhile they seek to preserve representative samples of wild nature (Hummel 1989). The **integrist** orientation is inherent in the concept of integrity of the combined natural/cultural ecosystem.

Integrity may have a two-tiered meaning whereby inherentists' ethical values are invoked but quasi-utilitarianist operations occur (Sagoff 1988). For me, the ecosystem concept, especially as applied to landscapes, provides an approximately level playing field for an evolving, adaptive process among the different major interests of the four types (RAAS 1989). I do not foresee a full synthesis and consensus of these interests in the near future. But we do want to do better than simply serve the lowest common denominator when used in a pejorative sense — whatever that really means!

No conventional political ideology within the Canadian and American parts of the Great Lakes Basin (or elsewhere in the western world) now has an ethical and conceptual base that relates clearly to an emerging vision of "ecosystem integrity." Exaggeration of the role of purely selfish behaviour, as in the theory of the "invisible hand" with respect to neo-classical free-market processes, predisposes a culture to behave disintegratively within itself, towards other cultures and towards non-human nature. As an allocative institution, the free market may be a lesser evil than the "invisible mind" of an hierarchic, centralized socialistic bureaucracy, but the latter is not the only alternative. Both of these allocative mechanisms have run wild during the 20th Century and both need to be domesticated.

Some opinion leaders forecast an integration in the 1990s of a number of North American reforms of the 1960s that have gone separate ways, i.e., reforms related to gender, race, the poor, war and the environment. In each of these reforms, ethical principles — of what is inherently proper and improper with respect to the rights of individuals, peoples or species — played important roles. Thus all are related to the "inherentist" value set as described above. Each of these reforms has had to contend with the tyrannies of the "invisible hand" and of the "invisible mind." Both the free market and the centralized bureaucracy are recognized as important and valuable cultural institutions; but, it is crucial that their roles be limited to things that are not valued highly within the culture. Each is prone to create "social traps" such as self-serving, self-reinforcing processes (Costanza 1987, Grima 1990) that are subversive of cultural integrity. Vigilance is necessary to spot and inactivate these traps as they first emerge. Otherwise, the market or the bureaucracy becomes society's master rather than its servant.

It may be that an integration of 1960s reforms in North America will create a basis for "green politics" here as has happened in Europe. A transnational version of green politics may now be emerging in the Great Lakes bioregion. **Cultural and natural integrity** may be an ethical watchword and an **ecosystems approach** may provide a relatively level playing field conceptually and methodologically (RAAS 1989).

With respect to the natural environment, a number of ethical concepts were written into U.S. federal laws in the early 1970s. One of those terms was "integrity" which was intended to refer both to the outcome of natural and of cultural self-integrative processes. Apparently, the main value orientation of these reformers was integrist as sketched above, with a strong leaning toward an inherentist position. Market-oriented utilist accounting of the benefits and costs of proposed measures to correct environmental degradation were expressly forbidden in some U.S. laws. This was apparently in response to practices by economic accountants over some decades in which *a posteriori* consequentialist values were quantified in monetary terms, and references to *a priori* deontological values were relegated to footnotes. By this means, the market-oriented utilists, with help from the exploitists, had influenced the political process.

Proponents of the term "integrity" for the U.S. federal laws (e.g., Woodwell, 1977, Jorling 1976, 1977) apparently understood that a political process in which benefit/cost ratios played a dominant role must threaten integrity, e.g., as that term was understood by Leopold (1949). The market place, and benefit/cost ratios in service of the market place, could be relied upon to safeguard and foster self-integrative processes neither in the natural nor in the cultural aspects of regional ecosystems. Though benefit/cost ratios might prove useful at some stage in a decision process, that stage was not to be at a high level of the decision process.

Similarly, a concept of "zero discharge" of hazardous pollutants has become law. Conventional utilists have argued that such a constraint is unrealistic. Reformers respond that the concept of zero discharge implies that no discharge can be legitimated *a priori* — any discharge however unintentional or accidental is morally wrong. Market-oriented utilists seek a numerical, monetary amount which the polluter might pay for a license to do such wrong, or to make good use of the assimilative capacity of the environment. But the reformers know that they have been down that path before — utilists have even suggested a number to be legitimated as a value for a human life.

In the Great Lakes Basin, reform principles have increasingly come to center on "integrity" as achievable within an "ecosystem approach." Conventional utilists, among the economists and engineers as well as among the conventional environmental and resource managers, have passed through stages of confusion, disbelief and anger. Some of them are now beginning to comprehend what cultural/natural integrity is coming to mean.

An Ecosystem in a State of Integrity

(James Kay and Bruce Bandurski collaborated with me on the following)

We know of no way to sketch the concept of ecosystem integrity in a linear, closed way. In fact, to do so would be contradictory to its meaning.

So we have sketched it as one long sequence of ideas, each partial. Thus an ecosystem with integrity:

- has as its primary nexus a set of living organisms, each unique but always changing, within adapting populations of different evolving species or taxa,
- with each organism being guided in part by genetic information some of which dates back billions of years,
- but also guided in part by recent information coded physiologically and behaviourally,
- and with a capability for creativity;

- contains organisms, perhaps including humans, that purposely modify their surroundings,
- through alteration of the system's abiotic features and secondarily through effects on other biota that in turn alter its features,
- through use of other biota as food,
- but seldom to an extent to affect drastically the entire ecological system so as to impair severely its biotic, self-organizing capabilities;

- contains some relatively large, longer-lived plant and animal organisms in locales where externally-driven factors — natural and cultural — do not reach extreme values with respect to preconditions necessary for life and do not fluctuate excessively with respect to amplitude and frequency,
- and these organisms serve to cumulate and integrate the effects of many of the ecological system's phenomena,
- and in turn partially regulate, in a cybernetic sense, many other features of the system;

- exhibits complex trophic networks within which high intensity energy from outside the system is embodied chemically as "food" to be transmitted from one organism to another,
- starting with photosynthesis or chemosynthesis followed by a biotic sequence of partial digestive lysis and new synthesis,
- such that both energy and "information" are transmitted in food,
- with an increase per unit of biomass in information content and "embodied energy" invested in generating that information up the trophic network,
- and with a continual emission of low quality energy in the form of heat;

- exhibits interactions between organisms within and across nested levels or scales of organization and interactions with non-living environmental phenomena of somewhat comparable though simpler organization,

- within a complex, spatio-temporal domain or landscape that is constraining but not determining,
- so that the whole system exhibits relatively persistent structures (or slow process phenomena) overlain with transient, perhaps cyclical, processes (or rapidly changing structures);

- contains elements or holons and, especially, the organisms that associate and organize themselves in complex ways with a balanced set of positive and negative feedback mechanisms,
- with the holons interacting competitively to their own limited advantage and reciprocally to others' long term advantage to achieve a dynamic mutualistic compromise,
- under a survival-related constraint that organizational flexibility (e.g., redundancy of options through latent structures or pathways) be retained to cope with inevitable surprises,
- which surprises generally have simplifying effects in the short term but complicating effects in the long term;

- may manifest itself sequentially and directionally in a number of somewhat different organizational states,
- with relative dominance by different sets of species in the different states,
- and perhaps with non-linear and rapid transition from one state to another;

- has locales or Anlagen within its spatio-temporal domain where self-organizational processes are more intense than elsewhere,
- and more complex both with respect to biotic and abiotic phenomena,
- with the self-organizing abilities radiating or networking outwards from such centres to interrelate in diminished intensity to those of other centres in ways analogous to gravitational forces and also
- in ways that transcend or penetrate what may superficially appear to be sharp physical boundaries;

- interrelates dynamically, as a large scale holon, with adjacent ecosystems simultaneously at a number of levels or scales of organization,
- at a boundary that may be interdigitated in a complicated way or may appear as a gradient,
- such that the apparent boundary of a system does not usually strongly determine system behaviour;

- comprises nested hierarchies or holarchies of subsystems each of which exhibits variations of a generic type of similar organizational regime, e.g., the Bertalanffy complex of processes,
- with all the ecosystem's organizational processes driven by some external sources of energy, usually sunlight, that contain some of the kind of energy, or exergy, appropriate and related to material movement and cycling and to photosyn thesis necessary for living processes,
- so that the external energy is dissipated through a nested holarchical process in which the "elements" (actually processes from a Bertalanffian perspective), at different spatio-temporal scales, exhibit some self-similar fractal features;

- is a self-organizing dissipative system which may represent a compromise between the thermodynamic imperative of energy destruction (the 2nd Law of Thermodynamics) and the biological imperative of survival,
- with the latter reflected perhaps in a will to live and to sustain identity;
- and so on.

Conclusion

There is room for choice in the kinds of ecosystems with integrity that humans might prefer. In human-dominated ecosystems, it is really a matter of: "What kind of garden do we want? What kind of garden can we get?" (Clark 1986). Forecasts of future ecosystems are not possible, but some future imaging of preferred ones is. With "backcasting" as a guide for determining how to get from where we are towards where we would prefer to be, we might be able to cultivate the kind of ecosystems we want to be in, providing always that we remain flexible enough to adapt to the surprises inherent in the dynamics of any ecosystemic garden at a scale so much greater than the human.

REFERENCES

Clark, William C. 1986. Sustainable development of the biosphere: themes for a research program. In: W.C. Clark and R.E. Munn (eds.). *Sustainable Development of the Biosphere.* Cambridge University Press. 491 pp.

Costanza, R. 1987. Social traps and environmental policy. *Bioscience*, 37:407-412.

Davidson, M. 1983. Uncommon Sense: The Life and Thought of Ludwig von Bertalanffy (1901-1972), *Father of General Systems Theory*. J.P. Tarcher, Los Angeles.

Feyerabend, P.K. 1988. *Against Method*. Verso, London, England, 296 pp.

GLWQB. 1985. *Report to the Great Lakes Water Quality Board to the International Joint Commission*, Great Lakes Regional Office, Windsor, Ontario.

Grima, A.P. 1990. *Great Lakes Levels Management: Relaxing the "Policy Trap"*. Proceedings of the American Water Resources Association and the Canadian Water Resources Association, Toronto, Ontario, April 1990.

Harris, H.J., V.A. Harris, H.A. Regier and D.J. Rapport. 1988. Importance of the nearshore area for sustainable redevelopment in the Great Lakes with observations on the Baltic Sea. *Ambio*, 17:112-120.

Holling, C.S. 1986. The resilience of terrestrial ecosystems: local surprise and global change, pp. 292-317. In: W.C. Clark and R.E. Munn (eds.). *Sustainable Development of the Biosphere*. Cambridge University Press, Cambridge, U.K. vi + 491. pp.

Hummel, M. (ed.) 1989. *Endangered Spaces: The Future for Canada's Wilderness*. Key Porter Books, Toronto, Ontario 288 pp.

Jorling, T. 1976. *Incorporating Ecological Principles into Public Policy. Environmental Policy and Law*, 2(3):140-146.

Jorling, T. 1977. Incorporating ecological interpretation into basic statutes, pp. 9-14. In: R.K. Ballentine and L.J. Guarraia (eds.). *The Integrity of Water*. U.S. Government Printing Office, Washington, D.C. No. 055-001-01068-1. vi + 230 pp.

Kuhn, T. 1970. *The Structure of Scientific Revolutions*. 2nd Ed. University of Chicago Press, Chicago, Illinois. 210 pp.

Leopold, A. 1949. *Sand County Almanac*. Oxford University Press, Oxford, U.K. xiii + 226 pp.

Norton, B.G. 1989. Intergenerational equity and environmental decisions: a model using Rawls' veil of ignorance. *Ecological Economics*, 1:137-159.

RAAS. 1989. *Towards an Ecosystem Charter for the Great Lakes - St. Lawrence*. The Rawson Academy of Aquatic Science, Ottawa, Ontario, iv + 112 pp.

Rapport, D.J. and H.A. Regier. 1992. Disturbance and stress effects on ecological systems. In: B.C. Patten (ed.). *Complex Ecology: the Part-Whole Relation in Ecosystems*. Prentice-Hall, New York. (In press).

Regier, H.A. and A.L. Hamilton. 1990. Towards ecosystem integrity in the Great Lakes-St. Lawrence River Basin, pp. 182-196. In: J.O. Saunders (ed.). *The Legal Challenge of Sustainable Development*. Canadian Institute of Resources Law, Calgary, Alberta, x + 401 pp.

Regier, H.A., P. Tuunainen, Z. Russek and L.E. Persson. 1988. Rehabilitative redevelopment of the fish and fisheries of the Baltic Sea and the Great Lakes. *Ambio*, 17:121-130.

Sagoff, M. 1988. *The Economy of the Earth*. Cambridge University Press, Cambridge, U.K., x + 271 pp.

Ulanowicz, R.E. 1986. *Growth and Development: Ecosystems Phenomenology*. Springer-Verlag, New York, xiv + 203 pp.

Van Hise, C.R. 1911. *The Conservation of Natural Resources in the United States*. The MacMillan Company, New York, U.S.A. xi + 413 pp.

Woodwell, G.M. 1977. Biological integrity - 1975. pp. 141-148. In: R.K. Ballentine and L.J. Guarraia (eds.). *The Integrity of Water*. U.S. Government Printing Office, Washington, D.C. No. 055-001-01068-1. vi + 230 pp.

Considerations of Scale and Hierarchy

ANTHONY W. KING
Environmental Sciences Division, Oak Ridge National Laboratory
Oak Ridge, Tennessee

Introduction

Explicit considerations of scale are increasingly a part of the process by which ecologists approach a variety of ecological issues and problems, and hierarchy theory is commonly used to address questions of scale. Thus, it is expected and appropriate that the evolving issues of monitoring and managing ecosystem integrity be concerned with questions of scale and hierarchy.

How do considerations of scale and hierarchical organization influence the measurement or management of ecosystem integrity? What considerations of spatial and temporal scale should go into the design of measures for ecosystem integrity? Answers to these questions first require some definitions. What is ecosystem integrity? What is an ecosystem? And what are scale and hierarchical organization in these ecosystems?

The Ecosystem Concept

Ecosystem As Entity

The concept of ecosystem is both widely understood and "diffuse and ambiguous" (O'Neill *et al.* 1986). The ecosystem may be specific or generic, referring to a particular ecosystem or some ecosystem type. There are references to the Cedar Bog Lake ecosystem (Williams 1971), the Isle Royale National Park ecosystem (Rykiel and Kuenzel 1971), Great Lakes ecosystems (Magnuson *et al.* 1980), the Hudson River ecosystem (Limburg *et al.* 1986), the Serengeti ecosystem (Sinclair and Norton-Griffiths 1979), southeastern [United States] ecosystems (Howell *et al.* 1975), forest ecosystems (Reichle 1981), tropical rain forest ecosystems (Golley 1983), an oak ecosystem (Zak and Pregitzer 1990), and a *Populus tremuloides* ecosystem (Ruark and Bockheim 1988). The ambiguity in this use of ecosystem is understood, and *ecosystem* is a useful handle for referring to "that ecological stuff out there, over there."

Ecosystem is variously defined as the collection of all the organisms and environments in a single location (Tansley 1935, McNaughton and Wolf 1979), any organizational unit, including one or more living entities, through which there is a transfer and processing of energy and matter (Evans 1956), or a system, i.e., a collection of interacting components and their interactions, that includes ecological or biological components (Lindeman 1942, Odum 1971, 1983, Golley 1983). Interactions in these latter ecosystems are through transfers of energy and matter (Odum 1983). Common to all these definitions, and at least implicit in the general usage described above, is the idea that the ecosystem includes the physical or abiotic environment in addition to biological components (e.g., organisms). Inclusion of the physical environment distinguishes the concept of ecosystem from that of community. Community generally refers to the assemblage of species or populations in a location without explicit reference to their physical environment.

Strongly associated with the concept of ecosystem is the concept of *ecosystem function*. Ecosystem function generally refers to the functioning or operation of the ecosystem, its integrated holistic dynamics, and not the role or job of the ecosystem. The distinction is analogous to the difference between the functioning of an automobile and its function as a means of transportation. Ecosystem function is commonly associated with the dynamics of matter and energy processing and transfer. Biomass production and nutrient cycling, for example, are often referred to as ecosystem functions.

The concept of ecosystem function is implicit in the general use of ecosystem to refer to the collective ecology of a given location. There is a tendency to include the term ecosystem when attention is given to dynamics of biota-biota or biota-environment interactions and not just simply the area's biotic composition or structure. Consider the reference to *forest ecosystem*, for example, rather than simply *forest*. Minimally, use of ecosystem in this context implies a consideration of both the biology and physical (abiotic) environment of an area and a consideration of dynamic interactions among biota and between biota and environment.

Ecosystem structure commonly refers to the distribution of matter and energy among system components. Because of the emphasis on function, ecosystem components are frequently defined by their functional roles, especially their rate controlling attributes. The structural components may, for example, be biota-environment aggregates with common turnover times or rates of matter-energy processing. Ecosystem carbon pools may be distinguished by turnover times (e.g., rates of decay) without separating dead organic matter from the microbial populations that feed on it.

Descriptions of ecosystem structure frequently do not consider the distribution of matter or energy among populations or species. Normally (perhaps mistakenly from the perspective of system science) a middle ground is taken in which biological components are grouped *a priori* according to a mix of criteria including functional roles, morphological types, distributions in time and space, and occasionally coarse taxonomic distinctions. For example, in describing the carbon structure of a forest ecosystem, autotrophs (photosynthetic plants) may be distinguished from heterotrophs. The autotrophs may be further divided into trees and non-trees of overstory and understory. The trees may be yet further divided into deciduous and evergreen forms. Adding function to this structure requires an assignment of carbon transfers and turnover times to these otherwise defined components. Ideas of structure precede those of function. But even in these cases, the concept of ecosystem function often influences the choice of components. Deciduous and evergreen trees are distinguished rather than angiosperms and gymnosperms. Foliage morphology and behaviour influence ecosystem carbon flux, and there is sufficient overlap in these traits between the two broad taxonomic groups to necessitate the alternative differentiation by lifeform.

Ecosystem structure associated with matter and energy transfer rarely involves even broad taxonomic distinctions much less Latin binomials. There are, however, important exceptions to this generality. The individual-based models of biomass growth and succession in forests are good examples (Botkin *et al.* 1972, Shugart and West 1977, Shugart 1984, Post and Pastor 1990). These models distinguish individuals by species, but they also assign species (individuals) to functional types defined by their response to the physical environment (e.g., shade or drought tolerance; Shugart 1984). In general, the greater the focus on biotic interactions or dynamics like competition or species turnover (e.g., succession), the greater the tendency to distinguish ecosystem components by species nomenclatures. The individual organism-based models of forest growth explicitly treat competition between individual trees. They also consider the consequences of these interactions for successional changes in ecosystem attributes generally associated with ecosystem function (e.g., primary productivity and turnover in soil organic matter and nutrients; Pastor and Post 1986, 1988). This duality is reflected in the models' mixed distinction of ecosystem components by taxonomic and functional criteria.

Ecosystem as Perspective

A distinction can be made between a population-community approach to ecosystems and a process-functional approach (O'Neill *et al.* 1986). The former emphasizes species populations and interactions among them like competition and predation. The latter emphasizes the transfer and processing of matter and energy. In the population-community approach, the physical environment is seen as external to the system of biota and biotic interactions. In the process-functional approach the environment is an integral part of the system. In the extreme this dichotomy emerges as a distinction between community — the system of populations — and ecosystem—the system of matter-energy transformations through biota and environment. The separation of community and ecosystem ecology in textbooks and classrooms is evidence of the acceptance of this dichotomy.

Note, however, that both approaches emphasize interactions. One emphasizes biotic interactions, the other fluxes of matter and energy. Thus *ecosystem* may be identified as a perspective, a particular way of looking at the biota and environment of an area. *Community* is a different perspective. Alan and Hoekstra (1990) refer to these perspectives as criteria for distinguishing foreground from background, and for distinguishing an object from its context. The ecosystem criteria distinguishes fluxes of material and energy; the community criteria distinguishes collections of species. Allen *et al.* (1984) discuss the observer's choice of phenomena, the structurally defined entities that are designated as significant. In systems science these perspectives are criteria for system specification. They are criteria for the definition of observables (Rosen 1977). The investigator chooses to order, describe, and observe the system of biota and environment according to one of these perspectives. The ecosystem is not ontologically different from the community, they are simply different ways of looking at the same "stuff."

Ecosystem as a Hierarchical Level

The ecosystem is often identified as one level in an ecological or biological hierarchy extending from cells to biosphere (e.g., Odum 1971, Krebs 1978). This "traditional" (O'Neill *et al.* 1986) or "conventional" (Allen and Hoekstra 1990) hierarchy is an ordering based on structural organization. A collection of entities at one level is organized to form the next higher level. Cells are organized as organs, organs as organisms, organisms as populations, and populations as communities. According to this concept of ecosystem, the ecosystem is a higher level of organization than the community, and ecosystems contain communities just as organisms contain cells. The concept of ecosystem as level is pervasive, and there is frequent reference to the ecosystem level (Rapport *et al.* 1985) or ecosystem-level properties (Rapport *et al.* 1985, Huston *et al.* 1988, Clark 1990).

The concept of ecosystem as hierarchical level is counter to the concept of ecosystem as perspective or observational criterion. In the perspective concept, ecosystems are not composed of communities. Ecosystem and community are instead complementary descriptions of the same ecological system (O'Neill *et al.* 1986, Allen and Hoekstra 1990).

A consequence of the traditional hierarchical view of ecosystems is the idea that ecosystems are larger than the communities they contain, just as organisms must be larger than the cells they contain. It is generally recognized that an ecosystem may be large or small (e.g., there is usually no objection to referring to the cow-rumen ecosystem), but the generality emerges that ecosystems are larger scale than communities or populations. By this argument, ecosystem (ecosystem-level) processes or properties are larger scale than community processes or properties. This idea has inadvertently been reinforced by recent considerations of scale and hierarchy in ecological systems (Allen and Starr 1982, O'Neill *et al.* 1986). These discussions generally assert that higher levels of hierarchical organization are larger (or coarser) scaled than lower levels. Higher levels occupy larger spatial extents and are characterized by larger time constants (e.g., slower rates or longer turnover times). Starting with the idea that the traditional ecosystem level is a higher level than the community level, it would seem to follow ecosystems and ecosystem processes are larger scale than communities and community processes.

Allen and Hoekstra (1990) caution against confusing conventional levels of organization with hierarchical levels defined by considerations of time and space scales. O'Neill *et al.* (1986) similarly caution against confusing traditional levels of organization with hierarchical levels distinguished by differences in rates. They point to the paradox generated by recognizing that traditional community processes, like succession, may be much slower than traditional ecosystem processes like nutrient processing. Similarly, the spatial scales of nutrient cycling processes (e.g., microbial decomposition) may be much smaller than the spatial scale of the community defined by a species-area curve (O'Neill *et al.* 1986). Allen and Hoekstra (1990) argue that the confusion and apparent paradoxes can be avoided by recognizing that the ecosystem is actually only one of several possible criteria for ordering observations across a range of spatial and temporal scales and not a scale-defined level. O'Neill *et al.* (1986) argue that observations on an ecosystem, from either the process-functional perspective or the population-community perspective, can be ordered by various criteria to form alternative, complementary hierarchical structures. In short, the concepts of ecosystem and ecosystem properties like ecosystem integrity are not limited to a particular hierarchical level or to particular space and time scales. An ecosystem exists across a range of scales and may include several hierarchical levels.

To summarize, an ecological system is a system description of the interacting biota and environment of some time-space domain (some place over some time period). Whether this system description distinguishes individuals or populations and interactions like competition (which transfer information among components) or functional components and the transfer of matter-energy is secondary to the primary concept that the ecosystem is a system. *Ecosystem* may be used as shorthand for ecological system (Odum 1983), but it should be remembered that for many, myself included, the term *ecosystem* invokes a biased view towards a system of matter or energy transfer, or a process-functional perspective. Nevertheless, in this chapter I will use *ecosystem* to refer to the most general notion of an ecological system occupying a particular place (and time), without regard to the specific criteria for system specification or ordering of observations.

Inclusion of the physical environment is a definitive element of the ecosystem concept. However, the environment may be represented by boundary conditions or external forces and does not necessarily have to be included as interactive system components. Thus, it is possible to refer to an ecosystem which includes the physical environment but, is defined by community criteria emphasizing populations and population interactions like competition.

The concept of *ecosystem* applies across hierarchical levels of organization and is not limited to the "ecosystem level" of the traditional hierarchy. Thus, an ecosystem may contain several hierarchical levels. Similarly, the concept of *ecosystem* is not limited to certain time or space scales (e.g., large scales), and an ecosystem may span a large range of spatial and temporal scales.

Ecosystem Integrity

Ecosystem Integrity As System Integrity

Integrity (excluding the notion of firm adherence to a code of conduct or behaviour) generally refers to the soundness or completeness of some thing, the state of being whole and unimpaired. The notion of ecosystem integrity is intuitively appealing and understandable. We wish our ecosystems to be sound, whole, and unimpaired, and we understand, intuitively, what it means for an ecosystem to be in that state. However, monitoring, managing, quantifying, analyzing, or legislating ecosystem integrity requires a more precisely defined, more objective or empirical, concept of ecosystem integrity. This is the aspect of ecosystem integrity that I address here.

Because an ecosystem is, I believe, first and foremost a system, it is proper to address ecosystem integrity from the perspective of system integrity. What then is the system's, and thus ecosystem's, state of being whole and unimpaired?

A system is defined both by its components and the interactions among them. Similarly, a system description simultaneously involves both structure and function—what are the components, how are they connected, and how do they operate together? System integrity thus implies the integrity of both system structure and function, a maintenance of system components, interactions among them, and the resultant behaviour or dynamic of the system (e.g., succession or the processing of energy).

Strictly speaking, loss of any system component or any change in interactions can be viewed as a loss of system integrity. The system is no longer whole; something is missing or displaced. Thus, the loss of even a single species or population (a structural component) could be viewed as a loss of ecosystem integrity. However, it appears to me that, to a considerable degree, the intuitive concept of ecosystem integrity is biased towards functional integrity, the state of being unimpaired. There is a bias towards the integrity of ecosystem function, a maintenance of the whole system's integrated dynamic.

Function is often a consequence of structure and therefore, a change in structure may alter function. Obviously, loss of all primary producers (a structure defined by function) has dire consequences for ecosystem function. Similarly, loss of a "keystone" species (e.g., Paine's [1966] starfish) can influence ecosystem function, and invasion by exotic species (an addition of structure) can, in some circumstances, alter ecosystem function (Vitousek 1986). However, many systems, including ecological systems, are amazingly resilient to alteration of structure. Whole system function is maintained despite the structural change. Changes in structure (components or interactions) may often have little, or very transient, impact on system function. For example, when system components are organized in parallel, loss of one or more components is often compensated for by a redirection of flow through remaining parallel components. Parallel structure in ecosystems is related to the idea of functional redundancy or functional equivalence. Ecosystem components (e.g., species) may perform equivalent functions (i.e., operate in parallel), and loss of one or more may produce very little change in whole system function (O'Neill *et al.* 1986, Vitousek 1986). Primary productivity or nutrient cycling may, for example, remain relatively constant while species composition changes (Harcombe 1977, Rapport *et al.* 1985) or dominant species are removed (Foster *et al.* 1980).

Systems may also possess more active mechanisms of resilience. Feedback loops in interactions among system components may compensate for structural changes in such a way that whole system function is maintained or quickly restored. Systems may thus show adaptation to structural change or even exhibit healing or recuperative powers. These kinds of responses are widespread and their existence readily accepted in clearly homeostatic systems like organisms (indeed the responses may be definitive). Ecosystems, may exhibit similar responses (Rapport *et al.* 1985), though some argue that these responses are only apparent or, at best, analogous to the responses of "truly" homeostatic systems.

The Dependency on Perspective

A change in structure, with little or no change in function, might be viewed as an insignificant loss of ecosystem integrity. If, however, the focus is primarily on ecosystem function, the change may not be considered a loss of ecosystem integrity at all. A classic example is a change in biodiversity (e.g., species richness) that produces no observable change in ecosystem function (e.g., primary productivity). Is the species loss an insignificant loss of integrity, or is it even appropriate to consider the change as a loss of integrity? Is there a loss of integrity if there is no consequence for ecosystem function?

Once again we are faced with the problem of perspective, the criteria for system identification and the ordering of observations. Changes in a system defined by one criteria may have little impact on observations of that same system defined by other criteria. Consider the extreme case posed by ecosystem perspectives involving aesthetic, ethical, or economic criteria. From these perspectives, the human observer is an integral part of the system. There is a flow of resource, value, or other currency between human and non-human components. The human is part of system function. The role may be passive, only receiving from the rest of the system (e.g., an aesthetic perspective), or the role may be active, receiving from and influencing system function (e.g., an economic or natural resource perspective). These perspectives can yield legitimate system descriptions, but they impart value to system components differently than more traditional scientific perspectives in which the human is not an integral part of the system. Translating ecosystem integrity defined from one perspective to notions of integrity for another can be problematic.

Forested ecosystems, for example, can recover from even extensive disturbance by fire or logging operations. The integrity of the system as a forest may be wholly retained (i.e., it returns to forest and not grassland or scrub) and species composition may change only slightly (although the distribution of biomass by species may be altered for some time). However, the aesthetic quality of the system may be severely damaged for much of the recovery period. The integrity of the system defined by "aesthetic interaction" with the biota has been lost. But has the integrity of the forest as a forest been compromised?

Consider also that rare species are often assigned the highest value or priority for preservation or use as indicator species. The attention given to rare species arises, in part, from the observations that rarity may be a consequence of declining populations in response to stress and rare species may be more at risk. These observations arise from a community perspective. There is also an aesthetic element. Humans are attracted to and value the rare or unique. Yet, in either case, because they are rare, rare species are unlikely to have much impact on ecosystem function (admitting the possibility of rare "keystone" species). The common species are more likely to be doing the brunt of the work in ecosystem function. Thus, while the persistence of rare (or endangered) species is a legitimate measure of integrity from a community (or aesthetic ethical) perspective, the persistence of common species may be more crucial to the ecosystem's functional persistence and integrity, and they may thus be more appropriate indicators of ecosystem functional integrity.

However, ecosystem function is often remarkably resilient to the loss of even common species. Witness the limited change in biomass dynamics of southern Appalachian forests following the demise the American chestnut, a formally common species (Shugart and West 1977, McCormick and Platt 1980). And what changes in North American ecosystems can be attributed to the loss of the once abundant carrier pigeon? Direct measures of functional properties (e.g., nutrient export [O'Neill *et al.* 1977]) may thus be more sensitive (better? appropriate?) measures of ecosystem functional integrity (O'Neill *et al.* 1977), but they, in turn, may be insensitive measures of the integrity of species composition.

Assessment of ecosystem integrity is strongly dependent upon the perspective from which observations are organized. Definitions and measures of ecosystem integrity from one perspective may complement, contradict, or be largely independent of those from other perspectives. Care must therefore be taken to define the perspective used in making statements about ecosystem integrity and in making inferences about integrity from other perspectives.

The strongest inference can be made by explicitly examining the integrity of alternative, complementary descriptions of an ecosystem. The work of Rapport *et al.* (1985) is a good example. Their recognition of a general ecosystem stress syndrome (i.e., a loss of integrity) includes indicators from both ecosystem and community perspectives (e.g., nutrient leaking and loss of biodiversity, respectively). But, even here, the perspectives are limited to those of "natural" ecosystems largely exclusive of the human component. Indicators of ecosystem integrity should include indicators from as many different perspectives and system descriptions, as practical. Those associated with human value judgements, like economics or aesthetics, should not be excluded by a prejudice for natural, ecological, or scientific perspectives.

Scale and Ecosystem Integrity

The scale of an ecosystem refers to the spatial and temporal dimensions of the ecosystem. How large an area does the ecosystem occupy, and over what time period does the system description pertain? Scale may also refer to the scale of the observation set used to define and describe the ecosystem. Here, scale includes concepts of both grain and extent (Allen *et al.* 1984, Turner and Gardner 1991). Grain is the finest level of temporal or spatial resolution in an observation set. The grain of an observation set determines the lower limit on how fine a distinction can be made with that observation set. Extent is the areal expanse or the length of time over which observations with a particular grain are made. Extent sets an upper limit on the distinctions that can be made with a particular observation set.

Specification of scale is a fundamental part of system definition (Levin 1975, Allen *et al.* 1984, O'Neill *et al.* 1986). The choice of scale, both grain and extent, with which a system is observed is a primary determinant of the resulting system description. Observations over one hectare and one year will lead to a different system description than observations over thousands of hectares and tens of years. Different extents encompass different components and interactions. Similarly, observations of different grain resolve different components, interactions, and dynamics. Once the scale of observation is chosen, the ensuing system description is largely determined. Consequently, those characteristics of ecosystem integrity which may be observed or inferred are largely determined.

The scale of an observation set used to define a system and measure ecosystem integrity may be determined by the scale of management units. I might for example wish to monitor or measure the integrity of the Great Smokey Mountains National Park ecosystem. My observations might then be limited to the spatial extent defined by the park boundary. It may be possible to construct a legitimate system description from observations within those boundaries, but the system description will be limited to the system existing over scales less than, or equal to, the extent of the management unit. I can make legitimate inferences about that system, but the limited extent of the observation set may not allow valid measurement or inference about those ecosystem attributes for which measures of integrity are truly desired and which may actually be attributes of a larger system. The extent of the observation set must be matched to the system attributes of interest. Specifically, the extent of the observation set must be larger than, or equal to, the extent of the system in question.

Systems require certain spatial and temporal extents for maintenance of system structure and function. A minimum extent may be required for some process to operate or interaction to take place. For example, gap-phase forest dynamics occur at the spatial and temporal scales of dominant canopy trees (Shugart 1984). Similarly, the trophic interactions of wolves, moose, and vegetation on Isle Royale are played out over particular space and time scales. Failure to observe the system at these scales can obscure system structure and function and make inferences about ecosystem integrity difficult or impossible. Furthermore, restricting the system to an extent less than the minimum required for interactions to occur can impact system function and may lead to a loss of ecosystem integrity. Fences may physically impede the flow of interactions in the spatially distributed system (witness the impact of agriculture and fencing on African steppe ecosystems) and management units (e.g., park boundaries, state lines) may isolate the influence of management practices to scales less than sufficient to maintain system integrity. Clearly, the area required to manage wolf populations can be much larger than that needed to manage the persistence of an endangered bog plant, and it may exceed the boundaries of politically defined management units. The extent of the observation set required to measure the integrity of the ecosystem supporting North American waterfowl populations is larger than the extent of any single management unit. The example of innovative management scales and practices required for the management of migratory waterfowl populations can be extended to other ecosystem components.

Ecosystem integrity is a scale-dependent concept. Maintenance of ecosystem integrity implies maintenance of some normal state or norm of operation (e.g., homeostasis or homeorhesis). Measuring or observing ecosystem integrity, or its loss, thus requires observations over sufficient temporal extent to identify and characterize this normalcy. We are prisoners of perspective, and our concept of normal is empirically bound to the scales with which we observe a system. Long term observations may reveal slow changes in a system component identified as constant with short term observations. Similarly, observations over a large area can reveal heterogeneity imperceptible from limited, local observations. Concepts of normalcy, constancy, variability, and thus, ecosystem integrity are only meaningful within bounds set by the scale of observation.

Fire or other disturbance, for example, is often revealed as a normal part of ecosystem operation and seen as necessary for the maintenance of ecosystem integrity when the system is viewed from a long term, large scale, perspective. Locally and in the near term, fire can thoroughly destroy all integrity of system structure and function (e.g., change in canopy architecture, species composition, and productivity). But the persistence of the ecosystem, its larger scale integrity, may in fact depend on the recurrence of these catastrophic smaller scale losses of integrity (Vogl 1980). Similarly, observed changes in species composition might be seen as indicating a loss of ecosystem integrity until a larger scale, longer term perspective reveals that these changes are part of a natural sequence of succession.

Hierarchy and Ecosystem Integrity

In simplest terms, a hierarchy is a ranked ordering. It generally implies a ranking by authority or dependency, but there are also static hierarchies (e.g., a hierarchy of size). As it applies to systems, a hierarchy is a ranked ordering of interactions. Control hierarchies are orderings of control, which system components control and are controlled by others (e.g., pecking orders, military command structures, or robotic controls). Structural hierarchies are orderings of subsystems within systems which are themselves parts of systems (Walters 1971). The organization of cells within organs, within organisms and squads within platoons, within companies are examples of structural hierarchies. Structural hierarchies generally exhibit some characteristics of control hierarchies (e.g., cell behaviour is normally constrained or controlled by its inclusion in the organism), but a system element can exhibit control beyond control over its component parts. For example, the control exerted by the alpha male in a baboon troop does not require that the alpha male be built of subordinate individuals. Thus, the distinction is sometimes drawn between nested (structural) hierarchies and non-nested (control) hierarchies (Allen and Starr 1982).

Systems can be ordered into hierarchies according to various criteria (O'Neill *et al.* 1986). Observations may be ordered by differences in rates, by tangible components, by spatial discontinuities, by traditional levels of organization, or even by arbitrary characteristics (e.g., colour). These alternative criteria lead to alternative hierarchical structures. Many of the criteria converge on similar structure (the robust transformation of Allen *et al.* 1984), and it has been suggested that rates may be a fundamental way of decomposing hierarchical structure (O'Neill *et al.* 1986). However, concepts, expectations, or predictions developed for hierarchies defined by one ordering criteria do not necessarily, or normally, transfer to hierarchies defined by other criteria (Allen *et al.* 1984). One must be careful to identify the ordering criteria involved.

There has been considerable discussion of hierarchy theory as it applies to ecological systems (Overton 1972, 1975, Webster 1979, Shilov 1981, Allen and Starr 1982, Patten 1982, Allen *et al.* 1984, O'Neill *et al.* 1986). This is not the place to attempt a review of that material. Rather, I will discuss aspects of nested hierarchies ordered by difference in rates that are particularly germane to the present discussion of ecosystem integrity. Much of the following discussion is drawn from O'Neill *et al.* (1986).

In a nested hierarchical system ordered by differences in rates, a level is identified by a distinct cluster of similar rates. Higher levels are characterized by slower rates. As a consequence of the physical nesting of structure, higher levels of nested hierarchies also occupy larger spatial scales. Thus, levels of organization in a nested hierarchical system can be ordered according to their relative position in a coordinate space defined by time and space scales (Figure 1). This coincidence, between level of organization and time and space scales, is an important theme in considerations of ecological hierarchies (Delcourt *et al.* 1983, Urban *et al.* 1987, Delcourt and Delcourt 1988, Carpenter 1989, King *et al.* 1990). Oceanographers (and limnologists?) have historically used a comparable structure to order their considerations of system variability (e.g., Stommel diagrams [Stommel 1963], Monin *et al.* 1977). Atmospheric scientists and geographers have adopted a similar approach (Dickinson 1988, Meentemeyer 1989).

The time and space scales of hierarchical organization represent scales of interaction (Allen and Starr 1982). The distribution of interaction frequency and strength define component subsystems or holons within the whole system. Discontinuities in the interactions define subsystem (holon) boundaries (Allen and Starr 1982). Rapid or frequent interactions occur within holons. Slower rates occur between holons. For example, in social mammals, interactions among family members are stronger and more frequent than interactions with other members of the local society, and interactions among the troop or band are more frequent than interactions between bands.

When interactions are dominated by physical characteristics of the ecosystem, scales of interaction, subsystem (holon) boundaries, may be obvious. In the terminology of Allen and Starr (1982) or O'Neill *et al.* (1986), the holon (subsystem) becomes tangible. For example, the spatial extents and water retention times of a topographically defined watershed represents a "natural" scale in the system bound together by the flow of water (e.g., the cycling of a water soluble nutrient). The time and space scales of the watershed delineate a specific, well defined, tangible subsystem. Interactions among subsystems at this scale, for example, the combination of first-order watersheds to form second-order watersheds, produce the next higher-level subsystem with its on natural scale. Physical processes of turbulent mixing in oceans and larger lakes can similarly impress natural scales on system interactions (e.g., the concentrations of phytoplankton growth and subsequent trophic interactions in eddies and gyres). The physical limitation set by the penetration of light (the euphotic zone) is another example. However, the phenomenon of "natural" scales applies equally to less physically bound subsystems. For example, in the system defined by competitive interactions among trees in a mesic forest, Urban *et al.* (1987) recognize scales of germination, tree replacement, suc-

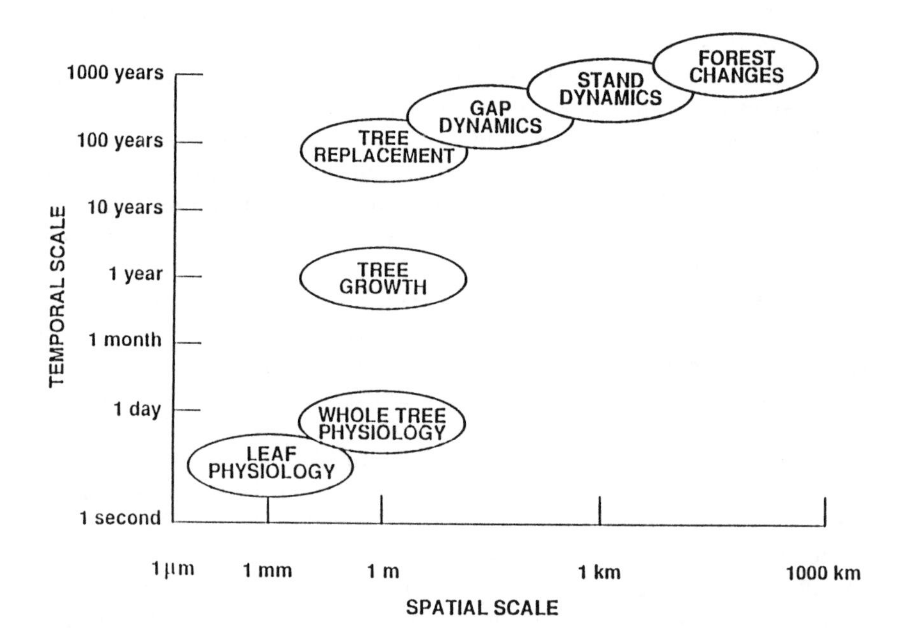

Figure 1. An example of the ordering of hierarchically organized rate dynamics by their associated spatial and temporal scales. The ellipses represent levels of carbon-biomass dynamics in a forested ecosystem. From King *et al.* (1990).

cession, species migration, and species extinction, in order of increasing spatial extent. Similarly, trophic interactions among wolves and moose define natural scales in that system. For example, the foraging area utilized by a wolf pack and the time required to cover that area define a scale for the individual pack. Interactions among wolf packs generate a next level of larger scale defined by the spatial and temporal distribution of the packs' foraging areas. In systems defined by human interactions with nonhuman components, natural scales may be defined by such things as areas of resource harvest or roadside vistas (e.g., how frequently humans see, interact with, different parts of a park).

In some cases, interactions may impress physical, observable patterns in space and time. For example, the scale of gap-phase dynamics (interactions among individual trees) may be evident in the spatial distribution of tree species and stand structure (Urban *et al.* 1987). The activity of a beaver colony can impress distinct patterns on the landscape at scales determined by foraging behaviour and social interactions (Johnston and Naiman 1987). These physical patterns or structures may reinforce the responsible interactions or influence interactions of other system components, even when viewed from alternative perspectives. Consider the impact beaver ponds or wallows created by alligators, buffalo, or other large animals can have on the distribution and interaction of other species, nutrient cycling, and energy flow.

Interactions that occur within a level are symmetric. The holons within a level operate at similar rates and can mutually affect the behaviour of others. Relations, interactions, between levels are asymmetric. O'Neill *et al.* (1986) identify these asymmetric relationships as *constraints* (O'Neill *et al.* 1989). *Constraint* is a very general term, simultaneously useful and potentially confusing. A molecule of gas may be constrained both by the walls of a container and by interactions with neighboring molecules. Here, I limit my use of *constraint* to those controls that emerge as a consequence of system interactions. Other extra-system controls (e.g., the walls of the container) I refer to as boundary conditions.

Constraints from lower levels appear as *potentials* (the biotic potentials of O'Neill *et al.* [1989]. Behavior of a system at one level is partially a consequence of the aggregate, integrated, behaviour of lower level components. These lower level behaviours can set limits on the behaviour of higher levels. They are particularly evident as rate-limiting interactions (analogous to rate-limiting steps in chemical dynamics). For example, a higher level cannot process material at a rate faster than the fastest rate of its functional components responsible for that processing. Salthe (1985) refers to these lower-level limitations as initiating conditions.

Within a hierarchical system, potentials are normally not realized. Interactions among system components at a given level constrain the expression of a behaviour to be less than the potential provided by lower levels. Again, with reference to rates, the behaviour will be slower than the potential. As noted above, it is this difference in rates that defines the levels of the hierarchy. The interactions at one level generate the components of the next higher level, forming the potential for those higher level components. The aggregate behaviour of components at one level is a property of the next higher level and appear to the individual components of a given level as part of their environment. O'Neill *et al.* (1989) thus identify constraints from higher levels as *environmental limits*. These limits may arise from the immediately higher level, as a consequence of symmetric interactions with components of similar rates, or they may arise from higher levels. The further removed, the slower the rates, the more likely they are to appear as part of a very slowly changing or constant environment. At a given level, a single component acting alone has little influence on the aggregate behaviour of the next higher level of which it is a part. However, it may be strongly constrained by that aggregate.

Consider a couple of simple examples. While it is true that a wildebeest herd migrates as a consequence of the movement of individuals, any single individual has very little influence on the stampeding behaviour of the migrating herd. The individual wildebeest is, however, strongly constrained to follow the herd, to "go with the flow." More subtly perhaps, a canopy is made up of individual leaves and canopy photosynthesis is a consequence of the photosynthesis of individual leaves. At the same time, the canopy is the environment of any individual leaf, and single leaf photosynthesis is constrained by this environment. Realized canopy photosynthesis is less than the sum of potential individual leaf photosynthesis.

The asymmetry of interaction in the vertical structure of a hierarchical system has important consequences for system integrity and response to perturbation. Signals or fluctuations from lower levels are attenuated, damped, by successively higher levels. The further removed the lower level is from the level being observed, the more complete the attenuation. The responses of higher levels are integrated, averaged, responses of lower levels. Thus, a detectable response of a higher level to perturbation, fluctuation, in a lower level requires either very strong change in one or a few lower level components or very extensive change in most or all of the lower level components. The change must perceptibly alter the average of the lower level components. This property of normal hierarchical structure isolates higher level organization from all but the most extreme or extensive lower level fluctuation or perturbation. When the higher levels begin to respond at frequencies very close to that of the lower levels, the hierarchical structure of the system has degraded. Whole system integrity is thus maintained by the asymmetric between-level relationships (or loose vertical coupling [Simon 1973]) of hierarchically organized systems.

Because system components at lower levels are constrained by the higher levels, changes or perturbations at higher levels frequently impact lower level components. Perturb a higher level and its component lower levels are largely constrained to follow. The reverse is not true. Most lower level perturbations will be attenuated. Consider again the wildebeest. Disruption of the wildebeests' annual migration route, an effort requiring large scale intervention, has an affect on every member of the herd. Removal of one individual, even a few, will have little or no consequence for the herd's behaviour. If we remove enough individuals, the herding behaviour or constraint is diminished by a diminution of interactions among individuals. The aggregate begins to act less-and-less like a herd, eventually collapsing into erratic individual behaviours. This loss of higher level organization will have dire consequences for the individual wildebeest dependent upon the context of the herd, and it can impact other components of the Serengeti ecosystem that are dependent upon the wildebeests' annual migration.

The relationship between scale and hierarchical level and the asymmetric relationship between levels suggest an expected relationship between the scale of perturbation and whole-ecosystem integrity. Local, fine scale, or short term perturbations are likely to be attenuated as they are transferred through paths of interaction to larger, longer term, coarser scales. The consequences of fine scale perturbations are seen in the integrated averaged behaviour at larger scales. This average will attenuate or mask all but the most extreme or extensive smaller scale perturbations. The greater the difference is between the scale of perturbation and the scale of observation for ecosystem integrity, the greater the degree of attenuation. This theoretical consideration reinforces the common sense expectation that localized short term perturbations are unlikely to have consequences for large ecosystems and long periods of time. When they do, chaotic behaviour (e.g., extreme sensitivity to initial conditions) or catastrophic behaviour (e.g., bifurcation) is an indication that: (a) the system is not organized as a nested, rate-ordered hierarchy, (b) the hierarchical structure has deteriorated, or (c) the system is in transition between alternative hierarchical organizations (O'Neill *et al.* 1989). In these circumstances, hierarchy theory will be of limited use in the analysis of ecosystem integrity.

Scale, Hierarchy, and Measures of Ecosystem Integrity

Ecosystem integrity is a holistic, whole-system property. It applies to the entire integrated system and not just one or more of its components. Core body temperature is similarly a holistic measure of thermal state in homeotherms, and a crude analogy can be drawn between measures of body temperature and measures of ecosystem integrity. However, most ecosystems at issue extend over large areas and persist for long periods of time. It is thus difficult to devise large scale, single-valued measurements of ecosystem integrity comparable to using a thermometer to measure body temperature. Whatever empirical measurements we ultimately employ to characterize ecosystem integrity, these measurements will almost invariably come from scales and levels of organization smaller, finer, or lower than the entire system. Remote sensing holds the promise of larger-scale ecosystem measurements, but effective use of these measurements will require the determination of meaningful relationships between spectral signatures and other ecosystem attributes reflecting ecosystem integrity. In the interim, and to complement these efforts, we must normally deal with a set of measurements from scales smaller than the whole system. From this set of measurements, we must devise some collective integrated measure for the aggregate system.

The scaled interactions of a hierarchically organized ecosystem can act as natural integrators of local processes. This natural integration is especially evident in scales defined by physical characteristics of the system. Watersheds are a good example. Water flowing through the watershed integrates many local changes over a rather clearly defined extent. Similarly, the foraging of a wolf pack integrates the finer scale temporal and spatial variability of their prey.

Advantage can be made of these natural scales of integration. Focusing a field measurement campaign on these natural integrators can reduce the number of local measurements required to characterize the entire system. Measuring output at the point of watershed discharge provides an integrated, holistic, measure for the entire system over the spatial extent of the watershed. Only a single point measurement at the weir is needed to characterize the entire spatial extent of the watershed. The weir measurement also integrates temporal variations over the period of the watersheds retention time. Successive measures at the weir over time provide for further integration. Thus, if I could identify a single watershed containing the entire ecosystem of interest, I would be tempted to focus my measurements there, at the point of discharge. In a sense, I can "take the temperature of the watershed." Airsheds and bodies of water with periodic turnover are related examples of natural physical integrators. Additional examples of a slightly different nature include tree rings, which integrate small scale variability in small scale individual leaf photosynthesis (O'Neill *et al.* 1986), and top level consumers, which frequently integrate toxic contamination of the ecosystem through the process of bioaccumulation.

There are also disadvantages in using these natural integrators. Because they integrate a diversity of natural and anthropogenic variations, it may be difficult to ascribe cause and effect to observed changes. Annual wood production and stream nutrient concentration are each influenced by many factors. Isolating a single cause (e.g., an anthropogenic disturbance) can be difficult. Witness the difficulty of assigning cause to observations of forest dieback at higher altitudes of the eastern United States. Observed mortality could be a consequence of acid precipitation, but forest growth (or lack of it) integrates so many other factors (including natural succession) that it is difficult to precisely define the contribution of any one factor.

Larger scale, natural integrators also tend to attenuate or smooth out finer scale variability. This attenuation tends to remove fine scale noise from observations and it makes detection of significant signal or trend at larger scales more likely. However, it also filters out finer scale signals that may be indicative of developing problems. We are caught in the dilemma of making observations at the appropriate larger scales of the system, but with the nagging concern that we may be missing significant signals from finer-scales and limiting our ability to ascribe cause-and-effect to our observations. In compromise, larger scale measurements, taking advantage of natural integrators, may need to be complemented with finer scale measurements.

Despite their disadvantages, natural scales of integration, if they can be identified, should be utilized in the sampling design for measures of ecosystem integrity. Given the practical limitations on making field measurements over extensive areas and for long periods of time, it must surely be appropriate to structure the sampling design for most efficient returns. When the system is characterized by natural scales of interaction or integration, this means sampling with respect to those scales. For example, an ecosystem will normally exist across a range of time and space scales. Once the time-space boundaries of the ecosystem are identified, either by consideration of process or the boundaries of management units, it is essential that the observations used to characterize ecosystem integrity characterize the entire extent of the ecosystem. Finer scale heterogeneity is one of the factors determining the nature of this sample. If, for example, the ecosystem were the same over the entire extent (both in time and space) with respect to the observations for ecosystem integrity, only a single, smaller scale measurement would be necessary to represent the entire extent. As the heterogeneity of the system increases, a larger sample of finer grain is required to effectively characterize the entire ecosystem. Targeting scales of integration as close to the scale of the entire ecosystem as possible will reduce the number of samples necessary to meet that requirement. For example, fewer second order watersheds must be sampled than first order watersheds to characterize a landscape. The identification of natural scales of integration may be used to locate holistic integrated measures like those at watershed weirs, or the scales may be used to stratify multiple, disaggregated, finer scale samples. These approaches might also be combined. Again using the watersheds as an example, a sampling strategy might take advantage of the scale structure of an ecosystem by including integrated measures at the weirs of ith-order watersheds (e.g., stream water chemistry) and supplementing these measurements with samples of soil columns spatially distributed within one or more of the ith-order watersheds. The distribution of samples within these watersheds would logically be stratified by watersheds of the ith-1 order. Similarly, holistic, integrated measures of the quality of wolf habitat in a wolf-moose ecosystem might be obtained by monitoring the fecundity and survivorship of all wolf packs within the extent of the ecosystem. Finer scale measurements of individual wolf health, prey density, and vegetation could be made within the home range of one or more of the packs.

Targeting scales close to the scale of the entire ecosystem of interest will also increase the likelihood that observed changes will be of consequence for the entire ecosystem. Changes in the soil chemistry of a few local sites are unlikely to be indications of changes in the chemistry of the entire ecosystem. Changes in the stream chemistry of higher order streams, on the other hand, are very likely an indication of a change in the entire ecosystem. Similarly, illness of a few wolves or local fluctuations in prey density will not likely be expressed as a change in the wolf population of the entire ecosystem. A decrease in reproductive output of even one pack is more likely to indicate a decline in the ability of the ecosystem to support wolves. Again, these larger scale integrated measures are invaluable in detecting change like a loss of ecosystem integrity, but they may have to be supplemented with finer scale measurements to determine cause and effect.

Natural scales of integration may be identified by considering the interactions that bind a system together and the medium of that interaction (e.g., nutrients and water flow or trophic interactions and the area traversed in foraging). Tangible physical attributes of the system may clearly demarcate scales of the system (e.g., watershed topography, windthrow areas, or lake shores and other physical barriers to movement). Quantitative analysis of spatial (or temporal) patterns may reveal less obvious scaled structure that can also be associated with hierarchical system dynamics (O'Neill *et al.* 1991). Allen and Starr (1982) discuss, at some length, methods of multivariate analysis of spatial and temporal time series data for identifying scale in community ecology. Turner *et al.* (1991) review the use of spatial statistics to identify scales in landscapes.

All of these methods and more may be utilized in identifying scales of integration in ecosystems. Decisions on which data to collect will primarily be influenced by the choice of indicators of ecosystem integrity (Rapport *et al.* 1985) which are, in turn, largely determined by the perspective used in defining the ecosystem. Allen and Starr (1982) note, with respect to the problem of deciding which data to collect and dealing with the bias of *a priori* perspective, that "The best we can do is to record as much as possible, as frequently as possible, for as long as possible, and hope that we have not innocently passed over crucial keys to understanding a system's behaviour." I believe this admittedly somewhat unsatisfying guideline applies to measures of ecosystem integrity. Allen and Starr (1982) go on to suggest that once the data are collected, data transformations and tools of data analysis are used to "illuminate the structure inherent in the data." Identification of scales of integration or hierarchical organization and other considerations of scale and hierarchy largely follow the choice of measures of ecosystem integrity, i.e., the perspective from which, or for which, integrity is being addressed.

It should be noted that the scales identified by any of the various analytical methods are as dependent upon the perspective or criteria used to order observations as are the chosen measures of ecosystem integrity. Furthermore, the scales of integration defined from one system perspective for one indicator may not coincide with those for another. For example, the scale of the ith order watershed subsystem in the system defined by waterflow may have little in common with the scale of the community or population bound together by competitive interactions or the dispersal of airborne propagules unfettered by local topography.

Because of the dependency of scales of integration on system definition, efforts should be made to identify the natural scale of the ecosystem from a variety of perspectives. This should at least be done for every perspective involved in the suite of indicators for ecosystem integrity. When natural scales of integration from different perspectives coincide, special attention can be given to measuring at those scales. These scales also tend to coincide with tangibles (e.g., watersheds) and they form natural targets for measuring or monitoring strategies. If, for example, a variety of perspectives converge on the scale of the ith order watersheds in an ecosystem, measurements for ecosystem integrity can be targeted towards these watersheds. Preservation of the integrity of these watershed subsystems may be crucial to the preservation of integrity for the entire ecosystem viewed from a variety of perspectives.

ACKNOWLEDGEMENTS

I thank Alan R. Johnson and Robert V. O'Neill for their critical review of this document. This chapter was produced while the author was supported by the Carbon Dioxide Research Program, Atmospheric and Climate Research Division, Office of Health and Environmental Research, U.S. Department of Energy, under contract DE-AC05-84OR21400 with Martin Marietta Energy Systems, Inc. Publication No. 3948, Environmental Sciences Division, Oak Ridge National Laboratory.

REFERENCES

Allen, T.F.H. and T.B. Starr. 1982. *Hierarchy: Perspectives for Ecological Complexity.* The University of Chicago Press, Chicago.

Allen, T.F.H., and T.W. Hoekstra. 1990. The confusion between scale-defined levels and conventional levels of organization in ecology. *J. Vegetation Science.* 1:5-12.

Allen, T.F.H., R.V. O'Neill, and T.W. Hoekstra. 1984. Interlevel relations in ecological research and management: some working principles from hierarchy theory. USDA Forest Service General Technical Report RM-110, 11 p. Rocky Mountain Forest and Range Experiment Station, Fort Collins, Colorado.

Botkin, D.B., J.F. Janak, and J.R. Wallis. 1972. Some ecological consequences of a computer model of forest growth. *J. Ecol.* 60:849-873.

Carpenter, S.R. 1989. Temporal variance in lake communities: blue-green algae and the trophic cascade. *Landscape Ecology.* 3:175-184.

Clark, J.S. 1990. Integration of ecological levels: individual plant growth, population mortality, and ecosystem processes. *J. of Ecology.* 78:275-299.

Dickinson, R.E. 1988. Atmospheric systems and global change. In: T. Rosswall, R. G. Woodmansee, and P. G. Risser (eds.), *Scales and Global Change.* SCOPE 35. John Wiley and Sons, Chichester, England.

Delcourt, H.R., P.A. Delcourt, and T. Webb III. 1983. Dynamic plant ecology: the spectrum of vegetation change in space and time. *Quat. Sci. Rev.* 1:153-175.

Delcourt, H.R., and P.A. Delcourt. 1988. Quaternary landscape ecology: Relevant scales in space and time. *Landscape Ecology.* 2:23-44.

Evans, F. C. 1956. Ecosystems as the basic unit in ecology. *Science.* 123: 1127-1128.

Foster, M.M., P.M. Vitousek, P.A. Randolph. 1980. The effect of *Ambrosia artemisiifolia* on nutrient cycling in a first-year old-field. *Am. Midl. Natur.* 103:106-113.

Golley, F.B. 1983. Introduction. pp. 1-8. In: F.B. Golley (ed.), *Tropical Rain Forest Ecosystems: Structure and Function.* Ecosystems of the World 14A. Elsevier Scientific Publishing Company, Amsterdam.

Harcombe, P.A. 1977. Nutrient accumulation by vegetation during the first year of recovery of a tropical forest ecosystem. pp. 437-378. In: J. Cairns, Jr., K. L. Dickson, and E. E. Herricks (eds.), *Recovery and Restoration of Damaged Ecosystems.* University Press of Virginia, Charlottesville.

Howell, F.G., J.B. Gentry, and M.H. Smith (eds.). 1975. *Mineral Cycling in Southeastern Ecosystems.* ERDA Symposium Series. U.S. Energy Research and Development Administration. CONF-740513. Available from National Technical Information Center, U.S. Department of Commerce, Springfield, Virginia.

Huston, M., D.L. DeAngelis, and W.M. Post. 1988. New computer models unify ecological theory. *BioScience.* 38:682-691.

Johnston, C.A., and R.J. Naiman. 1987. Boundary dynamics at the aquatic-terrestrial interface: the influence of beaver and geomorphology. *Landscape Ecology.* 1:47-57.

King, A.W., W.R. Emanuel, and R.V. O'Neill. 1990. Linking mechanistic models of tree physiology with models of forest dynamics: problems of temporal scale. pp. 241—248. In: *Process Modeling of Forest Growth Responses to Environmental Stress.* edited by R. K. Dixon, R. S. Meldahl, G. A. Ruark, and W. G. Warren. Timber Press, Portland, Oregon.

Krebs, C.J. 1978. *Ecology: The Experimental Analysis of Distribution and Abundance.* 2nd Ed. Harper and Row, New York.

Limburg, K.E., M.A. Moran, and W.H. McDowell. 1986. *The Hudson River Ecosystem*. Springer-Verlag, New York.

Lindeman, R.L. 1942. The trophic dynamic aspect of ecology. *Ecology.* 23:399-418.

Magnuson, J.J., H.A. Regier, W.J. Christie, and W.C. Sonzogni. 1980. To rehabilitate and restore Great Lakes ecosystems. pp. 95-112. In: J. Cairns, Jr. (ed.), *The Recovery Process in Damaged Ecosystems*. Ann Arbor Science, Ann Arbor, Michigan.

McCormick, J.F., and R.B. Platt. 1980. Recovery of an Appalachian forest following the chestnut blight or Catherine Keever—you were right. *Am. Midl. Natur.* 104:264-273.

McNaughton, S.J., and L.L. Wolf. 1979. *General Ecology*. Second Edition. Holt, Rinehart, and Winston, New York.

Meentemeyer, V. 1989. Geographical perspectives of space, time, and scale. *Landscape Ecology*. 3:163-175.

Monin, A.S., V.M. Kamenkovich, V.G. Kort. 1977. *Variability of the Oceans*. John Wiley and Sons, New York.

Odum, E.P. 1971. *Fundamentals of Ecology*. 3rd Edition. Saunders, Philadelphia.

Odum, H. T. 1983. *Systems Ecology: An Introduction*. John Wiley and Sons, New York.

O'Neill, R.V., B.S. Ausmus, D.R. Jackson, R.I. Van Hook, P. Van Voris, C. Washburne, and A.P. Watson. 1977. Monitoring terrestrial ecosystems by analysis of nutrient export. *Water Air Soil Pollut*. 8:271-277.

O'Neill, R.V., D.L. DeAngelis, J.B. Waide, and T.F.H. Allen. 1986. *A Hierarchical Concept of Ecosystems*. Princeton University Press, Princeton, New Jersey.

O'Neill, R.V., A.R. Johnson, and A.W. King. 1989. A hierarchical framework for the analysis of scale. *Landscape Ecology*. 3:193-205.

Overton, W.S. 1972. Toward a general model structure for forest ecosystems. In: J.F. Franklin (ed.), *Proc. Symp. on Research on Coniferous Forest Ecosystems*. Northwest Forest Range Station, Portland.

Overton, W.S. 1975. The ecosystem modeling approach in the coniferous forest biome. pp. 117-138. In: B. C. Patten (ed.), *Systems Analysis and Simulation in Ecology*. Vol. III. Academic Press, New York.

Paine, R.T. 1966. Food web complexity and species diversity. *Am. Nat.* 100:65-75.

Pastor, J., and W.M. Post. 1986. Influence of climate, soil moisture, and succession on forest carbon and nitrogen cycles. *Biogeochemistry.* 2:3-27.

Pastor, J., and W.M. Post. 1988. Response of northern forests to $C0_2$-induced climatic change: dependence on soil water and nitrogen availabilities. *Nature.* 34:55-58.

Patten, B.C. 1982. Environs: relativistic elementary particles for ecology. *American Naturalist.* 119:179-219.

Post, W. M., and J. Pastor. 1990. An individual-based forest ecosystem model for projecting forest response to nutrient cycling and climate changes. pp. 61-74. In: L. C. Wensel and G. S. Biging (eds.), *Forest Simulation Systems.* University of California.

Rapport, D.J., H.A. Regier, and T.C. Hutchinson. 1985. Ecosystem behaviour under stress. *American Naturalist.* 125:617-640.

Reichle, D.E. 1981. *Dynamic Properties of Forest Ecosystems.* Cambridge University Press, Cambridge, United Kingdom.

Rosen, R. 1977. Observation and biological systems. *Bulletin of Mathematical Biology.* 39:663-678.

Ruark, G.A., and J.G. Bockheim. 1988. Biomass, net primary production, and nutrient distribution for an age sequence of *Populus tremuloides* ecosystems. *Can. J. For. Res.* 18:435-443.

Rykiel, E.J., Jr., and N.T. Kuenzel. 1971. Analog computer models of "The Wolves of Isle Royale". pp. 513-541. In: B. C. Patten (ed.), *Systems Analysis and Simulation in Ecology.* Vol. 1. Academic Press, New York.

Salthe, S.N. 1985. *Evolving hierarchical systems: their structure and representation.* Columbia University Press, New York.

Shilov, I.A. 1981. The biosphere, levels of organization of life, and problems of ecology. *Soviet J. Ecol.* 12:1-6.

Shugart, H.H. 1984. *A Theory of Forest Dynamics: The Ecological Implications of Forest Succession Models*. Springer-Verlag, New York.

Shugart, H.H., Jr., and D.C. West. 1977. Development of an Appalachian deciduous forest succession model and its application to assessment of the impact of the chestnut blight. *J. Environ. Management*. 5:161-179.

Simon, H.A. 1973. The organization of complex systems. pp. 3-27. In: H.H. Pattee (ed.), *Hierarchy Theory*. Braziller, New York.

Sinclair, A.R.E., and M. Norton-Griffiths. 1979. *Serengeti: Dynamics of an Ecosystem*. University of Chicago Press, Chicago.

Stommel, H. 1963. Some thoughts about planning the Kuroshio survey. pp. 22-23. In: *Proceedings of Symposium on the Kuroshio, Tokyo.*

Tansley, A.G. 1935. The use and abuse of vegetational concepts and terms. *Ecology*. 16:284-307.

Turner, M.G., and R.H. Gardner. 1991. Quantitative methods in landscape ecology: an introduction. pp. 3-14. In: M.G. Turner and R.H. Gardner (eds.), *Quantitative Methods in Landscape Ecology*. Springer-Verlag, New York.

Urban, D.L., R.V. O'Neill, and H.H. Shugart, Jr. 1987. Landscape ecology. *Bioscience* . 37:119-27.

Vitousek, P.M. 1986. Biological invasions and ecosystem properties: can species make a difference? pp. 163-176. In: H. A. Mooney and J.A. Drake (ed.), *Ecology of Biological Invasions of North America and Hawaii.* Springer-Verlag, New York.

Vogl, R.J. 1980. The ecological factors that promote perturbation-dependent ecosystems. pp. 63-94. In: J. Cairns, Jr. (ed.), *The Recovery Process In Damaged Ecosystems*. Ann Arbor Science, Ann Arbor, Michigan.

Walters, C.J. 1971. Systems ecology: the systems approach and mathematical models in ecology. pp. 276-292. In: E.P. Odum, *Fundamentals of Ecology*. 3rd. Ed. W. B. Saunders, Philadelphia.

Webster, J.R. 1979. Hierarchical organization of ecosystems. In: *Theoretical Systems Ecology*. ed. E. Halfon, pp. 119-31. New York: Academic Press.

Williams, R.B. 1971. Computer simulation of energy flow in Cedar Bog Lake, Minnesota based on the classical studies of Lindeman. pp. 543-582. In: B.C. Patten (ed.), *Systems Analysis and Simulation in Ecology*. Vol. 1. Academic Press, New York.

Zak, D.R., and K.S. Pregitzer. 1990. Spatial and temporal variability of nitrogen cycling in northern Lower Michigan. *Forest Science*. 36:367-380.

Applying Notions of Ecological Integrity

ROBERT STEEDMAN and WOLFGANG HAIDER
Ontario Ministry of Natural Resources
Centre for Northern Forest Ecosystem Research, Thunder Bay, Ontario

Introduction

Two recent symposia have addressed the "health" or "ecological integrity" of large aquatic ecosystems. In 1988, the Great Lakes Fishery Commission and the Science Advisory Board of the International Joint Commission sponsored a workshop entitled *An Ecosystem Approach to the Integrity of the Great Lakes in Turbulent Times* (Edwards and Regier 1990). In 1989, a special session of the 32nd Conference on Great Lakes Research focused on *Fish Community Health: Monitoring and Assessment in Large Lakes* (Spigarelli 1990). These two volumes bring together a diverse and thoughtful group of papers addressing more or less exactly the objectives of our present essay.

It seems to us that the experts noted above have thoroughly explored the conceptual and practical scientific issues that are likely to be brought to bear on the subject of ecological integrity as applied to aquatic ecosystems. This is not to say that those experts have reached a useful consensus. The symposia show that there are significant differences of opinion as to the meaning and relevance of "ecological integrity," and that there is disagreement as to the measures that should be used to index it.

We feel, however, that, this disagreement is largely irrelevant to the practical and beneficial application of the concept. The relatively well developed body of theory and practice cited in Edwards and Regier (1990), Spigarelli (1990) and Hunsaker *et al.* (1990) offers a wide range of approaches to measure ecological integrity. In our opinion, lessons from these and other applications indicate that scientific management approaches remain ill equipped to deal with socio-economic or perceptual values of aquatic ecosystems.

Practitioners of ecological integrity frequently allude to the limitations of the expert management approach. For example, Edwards and Regier (1990), Ryder (1990), and Koonce (1990) all include references to the significance of socio-economic and other social values for making management systems successful. These limitations derive, from cultural and societal values, the trend towards public participation in the process of natural resource allocation (Lerner 1990), and the increasing demand for multiple use of natural resources. The importance of quantitative information about values and perceptions increases in degraded, intensely used environments. In these situations, the likelihood of restoring historical or pre-disturbance ecosystem states is low, and the potential relevance of social science to ecosystem design decisions is high, because of the range of activities available. We believe that an integrated management framework which combines the normative "scientific" aspects of ecological integrity with social perspectives on ecosystem health is desirable. Research on social values of natural resources, expressed in the form of attitudes, perceptions, and preferences of user groups or the public at large, is not new. However, few explicit links exist between this research and ensuing policies for several reasons:

- research in the natural sciences and the social sciences is conducted under divergent paradigms, for different purposes and with different motives,
- many experts and managers believe that they understand public needs, and therefore, include them in the decision making process intuitively and informally, and
- in most cases it is difficult to quantify opinions, perceptions, attitudes and preferences in the same way that biophysical structures, states and processes can be quantified.

If wise natural resource management requires the manager to integrate social and biophysical information, how can this be done? Traditionally, much of this decision making has relied on intuitive judgement and *ad hoc* decisions. We feel that the decision making process can be enhanced by amalgamating the paradigms of social science and ecology.

The objective of this essay therefore, is to outline some strengths and limitations of ecological integrity as a concept to assist monitoring and management of aquatic and terrestrial systems. To do this, we will briefly explore the scientific and cultural contexts within which various integrity measures or indices may be applied. Next, we will show that when adopting the concept of ecological integrity for practical ecosystem management, the "how" must always be determined by the "why," i.e., methodology is strongly driven by context. To help bridge the scientific and conceptual gap between expert-guided management and social-cultural relevance, we will outline the strategic choice model (SCM) as a tool to integrate information needs for socially relevant ecosystem management.

How to be Comfortable with the Concept of Ecological Integrity

Ecological integrity may be usefully addressed at a number of spatial scales and levels of organization, depending on information needs and practical analytical constraints. The various measurement approaches may also be incorporated into a general methodological perspective of ecological integrity (Figure 1, see also France 1990). An agenda to operationalize ecological integrity should include evaluation of four aspects of ecological integrity: bias, context, methodology, and method. The evaluation process might proceed as follows:

Step 1. What are the cultural biases and scientific assumptions underlying selection of standards for ecological integrity?

Step 2. What is the context within which information about ecological integrity will be used? Will the information be used to obtain or organize scientific knowledge of the ecosystem? Will it facilitate management or regulatory decisions related to ecological integrity? Will it facilitate communication with non experts regarding ecosystem health?

Step 3. What methodological basis will be used to compare observational data with the standards? Is it a simple comparison with regional standards or guidelines? Is it based on theoretical aspects of system behaviour? Is it an empirical comparison of different ecosystems, possibly on a regional basis? Is it a hybrid of two or more of these methodologies?

Step 4. What are the actual measures that will be used to provide information about ecological integrity? Are they large scale integrative measures, or small scale process measures? What is the balance between biophysical and perceptual or value-based measures?

Below, we expand briefly on these four elements.

Biases of Integrity

Cultural norms and perspectives pervade both the rationale to deal with environmental problems, and the science that is used to provide solutions. Methodological bias may be introduced through the particular expertise of the observer or agency undertaking the measurements.

BIASES	POLITICAL	CULTURAL	SCIENTIFIC NORMATIVE STANDARDS
CONTEXT	SCIENCE	MANAGEMENT	EDUCATION
METHODOLOGY	NORMATIVE	SYSTEMS—THEORETICAL	EMPIRICAL—COMPARATIVE
METHOD	Measurements Directed to Regulatory Standards or Guidelines CHEMISTRY BIOMASS SPP. RICHNESS PRODUCTIVITY PUBLIC HEALTH TOXICOLOGY	DIVERSITY STABILITY RESISTANCE RESILIENCE EFFICIENCY COMPLEXITY HARMONY ARTICULATION ASCENDANCY	IBI VEGETATION INDEX IMAGE ANALYSIS LAKE TROUT KEY INDICATOR SPECIES PARTICLE SIZE SCENIC BEAUTY CHOICE MODELS

Figure 1: Conceptual framework to guide formulation of ecological integrity measures.

In the construction and application of ecological integrity measures, the choice of standards is highly susceptible to bias. In the absence of a strong theoretical or empirical rationale, significant bias may be introduced through the selection of standards that reflect social or cultural values, e.g., high productivity, stability, large salmonids. The implications of bias should be apparent and subject to evaluation. Honest and objective practitioners should take considerable care in exploring and documenting their particular approaches to the practical problems of selecting standards.

The Context of Integrity

The context of ecological integrity, as we use it here, refers to the application or final use of a program to measure and monitor ecological integrity. Figure 1 includes contexts that span the most likely applications of the concept: science, management and education. Details of the conception, quantification and analysis of integrity will vary considerably among each of these three applications.

Ecological integrity may be addressed within a purely biophysical scientific framework, without reference to ethical, perceptual or societal constraints. In this context, quantitative measurement of integrity should be compatible with most activities undertaken by conventional ecological science, i.e., pattern seeking, explanation of process and mechanism, the ability to explain past and present ecosystem states, and the ability to predict future ecosystem states. Here, the emphasis should be on quantitative and unambiguous measures of ecosystem state, preferably calibrated to strong historical or theoretical standards for ecological health. This approach maybe most appropriate for relatively healthy and intact systems such as the Upper Great Lakes.

Applied contexts relating to environmental management will emphasize relevance and practicality co-equally with scientific rigor, and may be more flexible with regard to quantification. In many practical instances, the degree of quantification required for good ecological management decisions may be relatively low, perhaps as simple as "low-high," or "low-medium-high." In such applications, cost and speed of reliable assessment results may be more important than precise quantification.

In contexts relating to education and communication, the results of monitoring, science, or management activities need to be communicated to non experts (ranging from children to resource users to senior policy makers). The same conditions that apply to any form of science transfer also apply here — simple unambiguous concepts presented with relevant examples, using plain language. A strength of ecological integrity indices may be their ability to effectively communicate multi-attribute data to nonscientific audiences.

Methodology of Measurement

There are a number of ways to compare and evaluate observational data relating to ecological integrity. A normative approach will generate standards or norms for quantitative ecological indicators such as water quality, biomass, productivity, or biodiversity. Often, these standards will be derived from toxicological or epidemiological research and may be codified in policy or legislation. In the absence of such research, arbitrary, conservative values are often chosen to address practical needs for regulation. The assessment of ecological integrity is accomplished by direct numerical comparison of sample data with the standard or guideline. Data may be compared parameter by parameter, or as an index aggregating time or space.

Standards for high ecological integrity may also be derived from theoretical predictions about the behaviour of natural systems. There is a substantial body of theory dealing with ecosystem development, and with the response of ecosystems to disturbance (see for example Steedman and Regier 1987 for references relevant to this essay). Some inferences derived from this theory are amenable to quantification, and may be used to measure ecological integrity.

A third methodology involves empirical comparison of ecological attributes on a regional scale. In practice, this often involves selection of regional ecosystems believed to represent healthy, relatively undisturbed conditions. A suite of quantitative measures is then used to compare other systems with the high quality standard. The index of biotic integrity (Karr 1981) falls into this category, although the IBI also addresses a number of theoretical concepts (Steedman and Rieger 1990).

A Toolbox of Ecological Indicators

In aquatic biology there is a tradition of quantitative research that has addressed the need to measure degradation of rivers and lakes. Early forms included the various "Saprobiensystems" developed for European rivers (Hynes 1960), and other biotic indices based on the abundance and diversity of benthic aquatic biota such as bacteria, algae and invertebrates (Hynes 1970, Hilsenhoff 1982, Plafkin *et al.* 1989).

Indices based on biota avoid some of the problems of interpretability and lack of spatial and temporal integration associated with environmental measures such as water chemistry. This is particularly true when the biota are long lived and are of direct social, economic, or recreational interest, and provide information about energy sources and physical structure of the ecosystem, as is the case with many fish species. (Karr 1981, Ryder and Edwards 1985). However, there is considerable interest in the ability of small, short lived organisms such as phytoplankton and zooplankton, to provide early warning information about subtle or chronic aspects of ecosystem integrity, that may not be expressed in biota at higher trophic levels (Carpenter 1988, Ulanowicz 1986a, Schindler 1987).

The IBI, or Index of Biotic Integrity (Karr 1981, 1991 and this volume) has been widely applied as a measure of ecological integrity in aquatic systems. This index was developed to address the need to measure "biological integrity" as written into the 1972 Amendments to the U.S. Federal Water Pollution Control Act. The revised Great Lakes Water Quality Agreement of 1978, as amended in 1987, also uses "biological integrity" as a regulatory goal.

Other approaches to the measurement of ecological integrity are cited in Edwards and Rieger (1990) and Spigarelli (1990). Choice of the measure for a particular context and application will almost always be related to regional data sources and information needs (Hunsaker *et al.* 1990). Regardless of the technique chosen, good quality control is essential; independent observations and techniques should be used to quantify the response of the measure to gradients of environmental disturbance or degradation, and regional limitations to the usefulness of the measures should be documented.

For all these reasons, sound environmental management requires that aspects of ecological integrity, however defined, be integrated with user perception and/or satisfaction. Management practices can be implemented successfully only if the various user groups and interest groups affected support and accept the management approach.

The Strategic Choice Model (SCM)

An ecosystem can be considered a multi-attribute phenomenon, for which users and other interest groups have certain perceptions and preferences. Within multi-attribute preference research and choice modelling, two important distinctions need to be made:

(1) research can be based either on revealed or on stated preferences;
(2) within stated preference research, a fundamental distinction exists between compositional and decompositional approaches.

Revealed preference research infers preference from direct observation of behaviour, the physical manifestation of choice. Stated preference research, on the other hand, relies on individuals to express preferences or choice directly, i.e., in a survey.

In revealed preference models, utility functions are estimated from observed behaviour, based on the assumption that individuals use a utility-maximizing choice strategy. Such models have been criticized because, in addition to the underlying utilities/preferences, the estimated parameters are likely to be influenced by the availability of alternatives and the spatial characteristics of study areas (Louviere and Timmermans 1990b).

Hence, experimental methods (stated preference research) which are based on direct observation of preferences for alternatives were developed. Any type of experimental research which "...provides a quantitative measure of the relative importance of one attraction as opposed to another" has come to be known as "conjoint analysis" (Aaker and Day 1986). Among them, it is useful to distinguish between the trade-off approach (compositional approach) and the full-profile (decompositional) approach.

In the compositional approach, it is assumed that individuals can directly express the importance of each separate attribute and the relative and absolute position of each attribute of each alternative. Overall evaluations or preference scores are "composed" by the researcher who combines, according to some *a priori* assumed algebraic function, the measured importance weights and attribute positions.

In the decompositional approach, instead of measuring weights of attributes separately, the preference for complete attribute profiles (scenarios) is measured. The construction of such profiles is based on the principles of multivariate statistical experiments, i.e., fractional factorial designs. The resulting overall preference scores for such synthetically designed choice alternatives can be decomposed statistically into the weights and evaluations (utilities) of attribute positions (levels), given an *a priori* assumed combination rule that represents how individuals integrate the weights and attribute level utilities to form overall preferences or utilities (Louviere and Timmermans 1990c).

Under this approach respondents typically rate or rank the alternatives in a survey. In order to link these preference structures to actual choice behaviour, an additional modelling step is required. This problem has been overcome by using a second (fractional) factorial design for combining choice alternatives into choice sets and respondents are now required to actually choose one alternative from a predescribed set (Louviere and Woodworth 1983, Louviere 1988a and b).

The advantages of this strategic choice model (SCM) can be summed up as follows:

- researcher has control over the variables included,
- many variables can be included for simultaneous evaluation,
- variables are independent (orthogonal),
- parameter estimates are derived for each variable and each level
- interactions and cross-effects can be estimated, and
- complete scenarios, including non-existent alternatives can be evaluated.

Below, we present a simple example illustrating how the SCM could be operationalized in the context of aquatic systems in northern Ontario. Table 1 lists some variables relating to information that might be measured as part of a multivariate ecological integrity index such as the IBI. Here, these variables are formulated in a way that allows their inclusion in a survey questionnaire that could be directed to nonexpert users of aquatic ecosystems. The examples in Table 1 would elicit perceptions about species richness or biodiversity of large fish, water body size, evidence of toxic chemicals in game fish, and simple aspects of water quality. By combining one level of each variable at a time, a hypothetical scenario or choice alternative can be formed. Choice alternatives can then be combined to form choice sets. Table 2 represents one example of such a choice set. Usually one questionnaire would cover 10 - 15 such choice sets.

The SCM constitutes a powerful method to increase the knowledge about public perception towards issues in natural resource management. The salience of many variables and their levels is determined in the appropriate context and, at the same time, a set of management alternatives can be evaluated. The SCM has been developed in the field of market research where it has found wide application. Examples for applications in recreation (Louviere and Timmermans 1990a, Haider and Ewing 1990) and natural resource management (Anderson *et al.* 1990) also exist.

Conclusion

The mandate of environmental and natural resource agencies to manage for ecosystem integrity is being confirmed and strengthened in legislation and policy at regional, provincial/state and national levels. There is a strong need for science that allows quantification of ecosystem and landscape integrity, relative to theoretical or empirical standards, at a variety of spatial scales. Relevant concepts and techniques exist to do this. However, ecosystem management according to ecological principles alone, as defined and implemented by experts, may be insufficient. Increasing pressure to use natural resources for a variety of purposes,

Table 1: Hypothetical variables for a strategic choice model for aquatic systems in northern Ontario.

VARIABLE AQUATIC SYSTEM	LEVEL
No. of Fish Species	1 3 5 7+
No. of Harvestable Fish Species	1 2 4+
Kinds of Fish Species	warm water cold water cool water
Lake Size	no lake <50 ha 50-200 ha 200-500 ha >500 ha
Stream/River Width	no river <5 m 5-10 m 10-50 m >50 m
Contaminant Levels in Fish (restricted for cons.) sport fish consumption restrictions due to contaminants	no limits 1 fish/week 3 fish/week
Algae Bloom on Water Body	none some common

Table 2: Example of a hypothetical choice set based on Table 1.

VARIABLE	CA1	CA2	CA3
No. of Fish Species	2	5	4
No. of Harvestable Species	1	3	5
Kinds of Species	warm water	cold water	cold water
Lake Size	n/a	50-200 ha	<50 ha
Stream/River Width	5 m	n/a	n/a
Contaminant Levels (restr. for consumption)	no	1/week	3/week
Algae Bloom	none	common	none
Which of the Alternatives Would You Select,If Any?	❑	❑	❑

combined with the increasing democratization of resource allocation decisions, have made social values an important component in the management process. Our approach to natural resource assessment explicitly links ecological science with public perception and values. Our approach is based in part on the quantitative methodology of strategic choice modelling, which facilitates measurement and modelling of values and choice behaviour. Management decisions that incorporate multi-attribute information of this type are more likely to balance scientific and social aspects of ecological integrity.

REFERENCES

Aaker, David A. and George S. Day. 1986. *Marketing Research*. 2nd Edition. New York, Wiley.

Anderson, Donald A., Clynn Phillips and Timothy C. Krehbiel. 1990. Wyoming angler attitudes and preferences; application of strategic choice modelling. Univ. of Wyoming Dept. of Statistics and Wyoming Cooperative Fishery & Wildlife Research Unit. 103 pp.

Carpenter, S.R. (ed.). 1988. *Complex interactions in lake communities*. Springer-Verlag, New York, NY. 283 pp.

Edwards, C.J. and H.A. Regier, (ed.). 1990. An ecosystem approach to the integrity of the Great Lakes in turbulent times. *Great Lakes Fish. Comm. Spec. Pub.* 90-4. 299 pp.

France, Robert L. 1990. Theoretical framework for developing and operationalizing an index of Zoobenthos community integrity: application to biomonitoring with Zoobenthos communities in the Great Lakes, p. 169-193. In: C.J. Edwards and H.A. Regier (ed.). An ecosystem approach to the integrity of the Great Lakes in turbulent times. *Great Lakes Fish. Comm. Spec. Pub.* pp. 90-4.

Haider, W. and G.O. Ewing. 1990. A model of tourist choices of hypothetical caribbean destinations. *Leisure Sciences*. 12:33-47.

Hilsenhoff, W.L. 1982. Using a biotic index to evaluate water quality in streams. *Wisc. Dept. Nat. Res. Tech. Bull.* No. 132. 22 pp.

Hunsaker, C.T., D. Carpenter, and J. Messer. 1990. Ecological indicators for regional monitoring. *Bull. Ecol. Soc.* Vol. 71(3):165-172.

Hynes, H.B.N. 1960. The biology of polluted waters. Reprinted in 1974 by University of Toronto Press, Toronto, Ontario. 202 pp.

Hynes, H.B.N. 1970. The ecology of running waters. University of Toronto Press. Toronto, Ontario. 555 pp.

Karr, J.R. 1981. Assessment of biotic integrity using fish communities. *Fisheries*. 6(6):21-27.

Karr, J.R. 1991. Biological integrity: a long-neglected aspect of water resource management. *Ecological Applications*. 1(1):66-84.

Lerner, S. 1990. Values in integrity, p. 37-44. In: C.J. Edwards and H.A. Regier (ed.). An ecosystem approach to the integrity of the Great Lakes in turbulent times. *Great Lakes Fish. Comm. Spec. Pub.* pp 90-4.

Louviere, J.J. 1988a. *Analyzing decision making-metric conjoint analysis*. Sage Publications Inc. Newbury Park, CA. 95 pp.

Louviere, J.J. 1988b. An experiment design approach to the development of conjoint-based choice simulation systems with an application to forecasting future retirement migration destination choices, p. 325-355. In: Golledge, R.G. and H. Timmermans (eds.). *Behavioral modelling in geography and planning*. Groom Helm, New York, NY. 497 pp.

Louviere, J.J. and H. Timmermans. 1990a. Stated preference and choice models applied to recreation research: a review. *Leisure Sciences*, 12:9-32.

Louviere, J.J. and H. Timmermans. 1990b. Preference analysis choice modelling and demand forecasting. O'Leary, J.T., D.R. Fesenmaier, T. Brown, D. Stynes, and B. Driver (eds.) In: Proceedings of the national outdoor recreation trends symposium III, Indianapolis, Indiana. March 29-31, 1990. pp. 52-64.

Louviere, J.J. and H.J.P. Timmermans. 1990c. Using hierarchical information to model consumer responses to possible planning actions: recreation destination choice illustration. *Envir. and Planning*. 22:291-308.

Louviere, J.J. and G. Woodworth. 1983. Design and analysis of simulated consumer choice or allocation experiments: an approach based on aggregate data. *J. of Marketing Res.* 20:350-367.

Plafkin, J.L., M.T. Barbour, K.D. Porter, S.K. Gross, and R.M. Hughes. 1989. Rapid bioassessment protocols for use in streams and rivers: benthic macroinvertebrates and fish. U.S. Environmental Protection Agency. Washington, D.C. 161 pp.

Ryder, R.A. 1990. Ecosystem health, a human perception: definition, detection, and the dichotomous key. *J. Great Lakes Res.* 16(4): 619-624.

Ryder, R.A., and C.J. Edwards (ed.). 1985. A conceptual approach for the application of biological indicators of ecosystem quality in the Great Lakes basin. Report to the Great Lakes Science Advisory Board of the International Joint Commission. Windsor, Ontario. 169 pp.

Schindler, D.W. 1987. Detecting ecosystem responses to anthropogenic stress. *Can. J. Fish. Aquat. Sci.* 44 (Suppl. 1): pp. 6-25.

Spigarelli, Steven A. (Ed.) 1990. Fish and community health: monitoring and assessment in large lakes. *J. Great Lakes Res.* Vol. 16(4). 670 pp.

Steedman, R.J. and H.A Regier. 1987. Ecosystem science for the Great Lakes: perspectives on degradative and rehabilitative transformations. *Can. J. Fish. Aquat. Sci.* Vol. 44(II):95-103.

Ulanowicz, R.E. 1986. *Growth and development: ecosystems phenomenology.* Springer-Verlag, New York, NY. 203 pp.

Choosing Indicators of Ecosystem Integrity: Wetlands as a Model System

PAUL A. KEDDY, HAROLD T. LEE, and IRENE C. WISHEU
University of Ottawa, Ottawa, Ontario

Introduction

Defining ecosystem integrity and choosing indicators for it would be a relatively simple task if the science of ecology was able to provide us with simple, rigorous models for describing and predicting these states of ecosystems. Unfortunately, a major problem in modern ecology is that we do not know which state variables are important and which ones are not (Rigler 1982, Peters 1991). Neither do we have simple quantitative models to describe relationships among these state variables. Lewontin (1974) calls this the "agony of community ecology." Determining the important state variables and describing the quantitative relationships among them is a key step for escaping many of the sterile controversies which detract from progress in ecology (Keddy 1987). In contrast, in assessing human health, we already know that some state variables, like height and hair colour, are unimportant in assessing state of health whereas others, like blood pressure and heart rate, are important and diagnostic.

Given that we lack the necessary rigorous ecological models, we need (1) to define integrity in an operational way, (2) to select those state variables which indicate integrity and (3) to identify levels of those state variables which indicate integrity or lack thereof (Rapport *et al.* 1979, 1985, Karr 1987, 1991, Schaeffer *et al.* 1988, Rapport 1990, Ryder 1990, Evans *et al.* 1990, Environment Canada 1991, Hellawell 1991, Keddy 1991, Lubchenco *et al.* 1991). We also, (4) need a system of feedback so that as we monitor our success or failure, we can modify the indicators and their levels.

The difficulty in coming up with potential indicators of wetland integrity is illustrated by two huge symposium volumes which have just been published (Sharitz and Gibbons 1989, Patten 1990). In spite of their size, these two volumes show that wetland biologists are collecting fragments of information from a wide array of vegetation types without a unifying conceptual framework which could guide a search for ecosystem integrity.

Our purpose here is to focus on indicators of integrity arising out of our own experiences working in wetlands. Our approach is pragmatic, in that we will accept existing definitions of integrity (e.g., Karr 1991, this volume) and list some indicators of integrity which we believe to be of value in wetlands. This list is not complete, but we believe that by starting a list, we will encourage others to contribute new indicators, and modify those already on the list. Our guiding philosophy in selecting these indicators was the belief that they should measure habitat and community state variables at a macro scale (Keddy 1991); we assume that in most cases individual species will persist if quality and diversity of habitat is maintained. The long term goal is to have an agreed upon list of indicators, with a series of constraints stating in which kinds of wetlands each indicator is appropriate, and which values of that indicator are desired. We suggest that this goal could culminate in a guidebook to wetland indicators, and suggest some steps to reach this goal.

A Path for Indicator Development

There are three key steps to developing useful and ecologically meaningful indicators (Figure 1): selecting the appropriate state variables, setting critical limits, and testing the indicators by following them through monitoring programs.

1. Selecting state variables

In the past, indicators have been developed haphazardly, often reflecting the interests of specific user groups and value systems, rather than according to more broad scale ecological criteria. This history is reflected in the kinds of databases we currently have to work with (C.J. Keddy and McCrae 1989). As Tansley said in 1914 (long before the advent of computer controlled data recording devices): "The mere taking of an instrument in the field and recording of observations ... is no guarantee of scientific results." We suggest that the following criteria should guide indicator selection:

(1) Ecologically meaningful — closely related to maintenance of essential environmental processes (e.g., water level fluctuations) and ecosystem functions (e.g., primary production),
(2) Macro scale — indicate changes in entire communities rather than selected species,
(3) General — can be measured meaningfully on different community types,

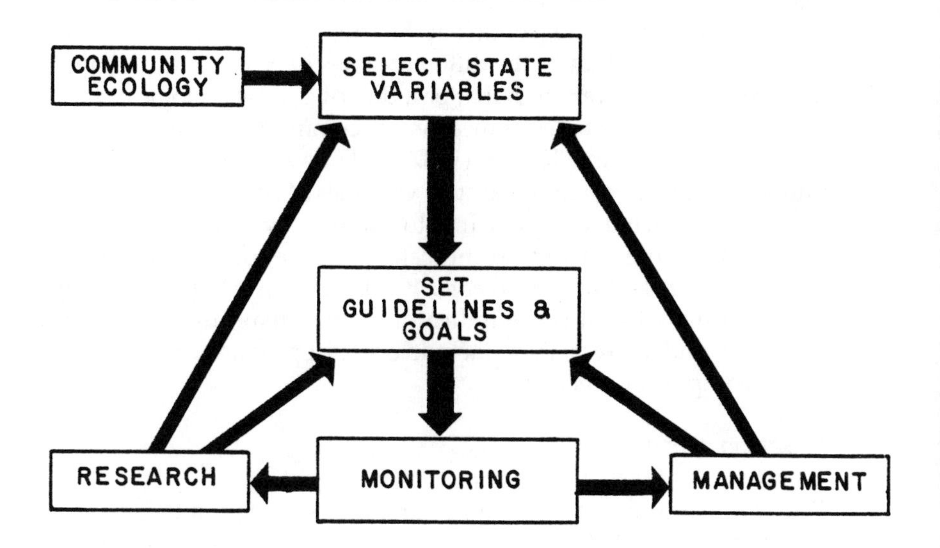

Figure 1: Framework for the identification, selection, and modification of indicators of ecosystem integrity.

(4) Sensitive — quickly respond to stresses and perturbations,
(5) Simple — are easy to measure.

With these criteria in mind, there are at least three ways in which we might proceed with selecting indicators.

First, we might begin with essential environmental factors which maintain and control the community type. Any ecosystem will have a large number of these factors, but it is usually possible to rank these in order of significance from those having large effects to those having smaller effects. For example, in wetlands, water levels and nutrient levels are very important factors. We might begin by selecting indicators which are sensitive to changes in these factors, and which have sufficient generality that they are not restricted to one species or one specific location.

A second approach is to look at the consequences of the selected environmental factors, that is, the different states of ecosystems that these factors create and the variables that best describe those biological states. We need to determine a series of measurable community state variables. Both Lewonton (1974) and Keddy (1987,1991) have discussed the strategic importance of determining the state variables of ecological systems.

A complementary approach is to examine systems which have been damaged or stressed to find indicators which appear sensitive to that damage or stress (Lubchenco *et al.* 1991). Rapport (1989) and Rapport *et al.* (1985) have listed some indicators which may identify ecosystems which are under stress; these include primary productivity, diversity, and size of component organisms.

2. Setting critical limits

Once we have selected our indicators, the next step is to set acceptable and desirable levels for them. That is to say, we must specify the level of integrity which we are trying to maintain. Each indicator would have a range of values specified, one limit being the tolerable level and the other being the desirable. If the system moved outside this specified range for given indicators, we would know that remedial action was needed to restore integrity. For example, we might set a goal of zero exotics as desirable for a rare wetland vegetation type, and two exotics as being tolerable. If more exotics than this invaded the site, we would investigate the reason for the invasion, and then take the appropriate remedial action.

What we envision, in the long run, is the development of a small handbook which would, 1) classify major ecosystem types, and 2) give for each ecosystem type the appropriate indicators of integrity with their desirable and acceptable levels. This would be a standard reference work not unlike those which guide foresters in determining ideal stand density for different forest and soil types or those which advise engineers in determining culvert sizes at stream crossings or which guide electricians in selecting wire gauge for different purposes. Figure 2 shows the structure such a guidebook might have, with a matrix of indicators by vegetation types. We imagine that some indicators (e.g., exotics) might have similar levels for all wetland types (e.g., swamp, bog, fen, marsh [Environment Canada 1987]), whereas others (e.g., amphibian biomass) might have different critical limits for different habitat types.

3. Monitoring

Selecting indicators and setting critical limits is obviously an evolutionary process. As our scientific knowledge of community ecology and our experience with ecosystem management increase, we need to be open to changing both indicators and critical limits. The handbook would therefore evolve to reflect our constantly improving knowledge of ecosystem integrity. It is essential to monitor after a major environmental project occurs, and then to use the information from monitoring to revise our criteria for future projects (Holling 1978, Beanland and Duinker 1983). A guidebook like this would therefore provide a focus for academic ecologists and those doing day to day monitoring, encouraging them to combine their cumulative experience.

Some Caveats

As nice as it would be, it is unlikely that all indicators will work across ecosystem types. For example, as Rapport *et al.* (1985) observe, there is little doubt that the loss of large organisms, in some systems, is an important indicator of ecosystem stress. When a forest is subjected to radiation, trees are the first plants to die (Woodwell and Rebuck 1967). However, in the wetlands which we have studied, there are some situations when invasion of large organisms actually indicates loss of integrity. The construction of the Bennett Dam demonstrated this with the modification of the flooding regime of the Peace-Athabasca River Delta leading to the invasion by woody plants. Keddy and Reznicek (1986) have developed a general model for wetlands that predicts that whenever the amplitude of water level fluctuations decreases, shrub invasion will result. Therefore, in marshes, the invasion of large plants is an indicator of loss of integrity. Shrub invasion is also taken to be a sign of stress in other vegetation types such as chalk grassland and prairie. We will therefore have to be careful that indicators developed to work at a coarse (i.e., global) scale, are not carelessly applied to the fine (i.e., local) scale. Thus we may not only need to distinguish, at times, among the four main wetland types (swamp, bog,

	COMMUNITY TYPES			
INDICATORS	SWAMP	MARSH	BOG	FEN
DIVERSITY				
GUILDS				
EXOTICS				
RARE SPECIES				
PLANT BIOMASS				
AMPHIBIAN BIOMASS				
↓	↓	↓	↓	↓
↓	↓	↓	↓	↓
↓	↓	↓	↓	↓

Figure 2: A possible structure for a guidebook to wetland indicators: A matrix of wetland indicators by the four wetland community types. Each cell would have a lower 'tolerable' and an upper 'desirable' indicator level.

marsh and fen), but we may need to distinguish at a finer scale yet, say between Great Lakes shoreline marshes, riverine marshes and prairie potholes.

Possible Indicators for Wetlands

Based on our experience with wetlands during the past decade, we offer the following preliminary list of six indicators for wetland integrity. For each we present one or more examples of their application. We emphasize that the critical limits for these indicators may vary among wetland types. Setting these critical limits would be the next step in developing the wetland integrity guidebook. The development of new indicators would proceed in parallel.

Diversity

There is a huge ecological literature on diversity, including quantitative measures for it and the implications of different levels of diversity for ecosystem function (e.g., Pielou 1975). Rather than take the scholarly route and reviewing many of the nuances of this literature, we will simply point out the common sense observation that all complex systems designed to persist through time have redundancy in them. For example, a spaceship has fall back electronic systems and often fallbacks to the fall backs. The term fall back comes from military history when a general would establish several lines of defense so that if one failed, one could literally fallback to the others. By analogy, any particular function in an ecosystem has levels of redundancy. For example, photosynthesis in a forest has a great deal of redundancy in that there are many species which carry out photosynthesis, so that if one is lost, there are others that may replace it. Similarly, within a single species, there are many individuals that carry out the similar function. Within an individual, there are many leaves and meristems. For this reason alone, we suggest that diversity is an essential component of integrity. In developing indicators of diversity, we should recognize that diversity is hierarchical just as photosynthetic ability is. Measuring species diversity may not be the most useful. Instead, one may wish to measure the diversity of guilds, that is, the number of kinds of functional groups.

Diversity was used as an indicator when we experimentally created 12 different wetland habitats replicated 10 times in a wetland mesocosm. All mesocosm cells were sown with a standard seed mixture. For each habitat type, 5 of the 10 replicates were subjected to eutrophication. Figure 3 shows that after two years, diversity had declined significantly in 8 of the

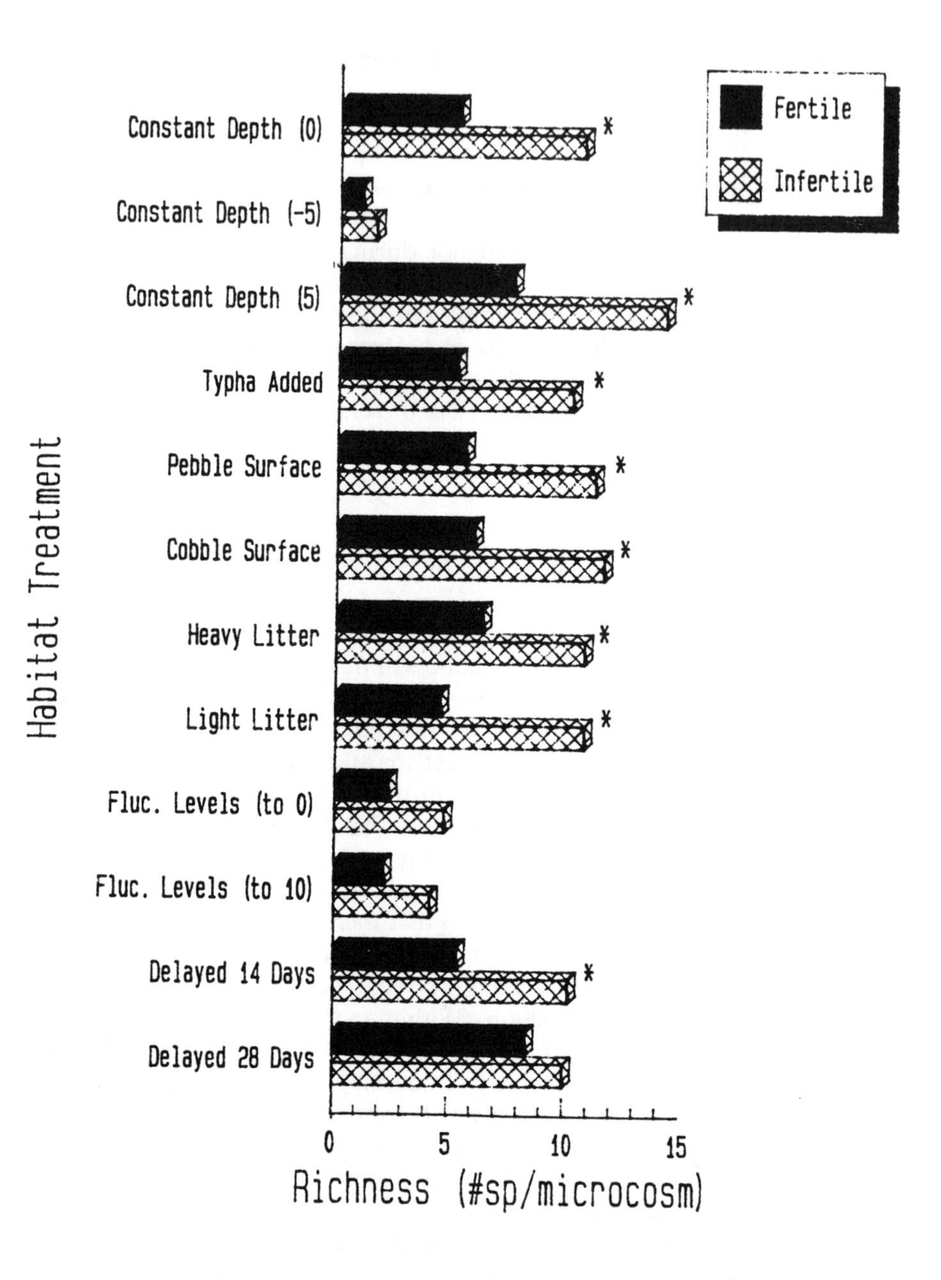

Figure 3: Species richness (no. of species / mesocosm) after two growing seasons for twelve artificially created wetland habitats at two levels of fertility. Significance at p=0.005 is indicated by a star (After Moore and Keddy 1989).

12 habitat types. Eutrophication is known to be a cause of stress in aquatic communities (Christie 1974) including wetlands (Moore 1990, Wisheu *et al.* in press) and thus threatens their integrity.

Indicator guilds

Guilds or functional groups of organisms are defined by their sharing certain traits which allow them to perform a similar function in a given ecosystem (Terborgh and Robinson 1986, Keddy 1990). Well known guilds in wetlands would include carnivorous plants, filter feeding invertebrates or dabbling ducks. The presence of a particular guild would be a good indicator that a particular function is being carried out or a good indicator that the habitat requirements of that particular guild were being maintained (Severinghaus 1981, Karr 1987).

Our own field experience in Atlantic coastal plain wetlands indicates that evergreen plants are good indicators for the habitat of other coastal plain species (Keddy and Wisheu 1989, Keddy 1981). Many evergreen species fall into a single plant guild, isoetids, which have a distinctive pincushion morphology and CAM photosynthesis (Boston and Adams 1987). Figure 4 plots the proportional abundance of the isoetid guild against the biomass of the vegetation of 121 wetland quadrats collected in Nova Scotia. Vegetation biomass is a good indicator of eutrophication (see Macroscale variables) so Figure 3 shows that the presence of isoetids is associated with infertile wetlands. The presence of isoetids therefore would be an indication that eutrophication was not a problem in a particular wetland. Here is where scale enters the picture however, since isoetids might be a excellent indicator of health in coastal plain wetlands, they would not be useful in indicating health in more fertile wetlands, like prairie potholes or Great Lakes wetlands on deep, fertile soils.

Exotics

Evidence of invading organisms will generally be an indicator of ecosystem distress. This could be because the invading organisms themselves are a symptom of a larger problem or because they themselves pose a threat where none existed before. Returning to the theme of enrichment in wetlands, Ehrenfeld (1983) has studied the impacts of human development in the wetlands of the New Jersey Pine Barrens. Nutrient enrichment is one of the principal consequences of human development. The percent of non-native flora increased from 12% at pristine sites to 96% at enriched sites (Table 1). The spread of Purple Loosestrife *(Lythrum salicaria)* is a another well documented example of an exotic which has had negative consequences for wetland integrity (Thompson *et al.* 1987).

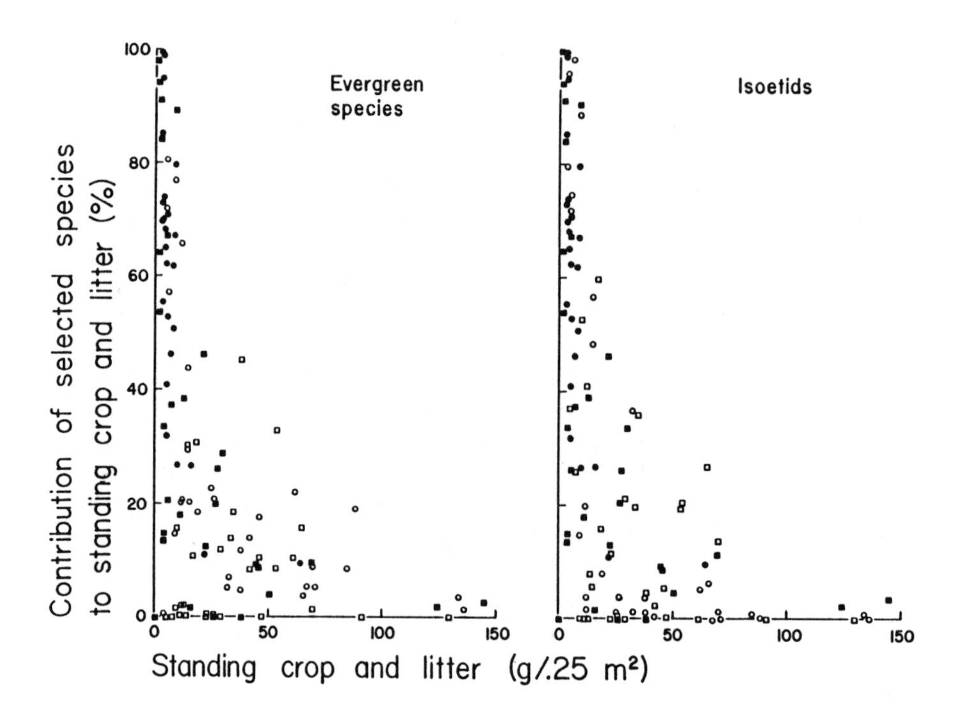

Figure 4: The contribution of evergreen and isoetid species to the standing crop and litter of each of 121 sites sampled from Wilson's Lake. The four habitats are represented as follows: high elevations on shores of red till (open circles), low elevations on shores of red till (closed circles), high elevations on shores of local till (open squares), low elevations on shores of local till (closed squares). (Reprinted from Wisheu and Keddy 1989 with permission from the publisher).

Table 1: Species occurring at enriched and pristine coastal plain sites in the New Jersey Pine Barrens (Ehrenfeld, 1983).

	No. of Species	% Carnivorous	% Non-native
Only at pristine sites	26	12	12
Only at enriched sites	72[1]	0	96

[1]Actual count was 73. One species was unidentifiable as to native or non-native.

Table 1 also amplifies two earlier themes. Although we presented evergreen species and isoetids as indicator guilds, Table 1 shows that carnivorous plants could also be used as an indicator of healthy wetlands in the Pine Barrens. Secondly, the Table shows how a single indicator, namely diversity, could be misleading. If we looked only at the number of species we would see that the enriched site had 3 times the number of species. Without the second indicator of percent exotics, we would fail to see that the higher number of species is actually an indicator of distress.

Rare species

The presence of a rare species is often an indicator of a system with integrity. At first this may seem counter intuitive but, in fact, most rare species are so designated because they have been found to be sensitive to human induced changes in their environments. The fact that rare species occur in a particular environment may therefore be an excellent indicator that human induced changes have been kept to a minimum (or else that the environment duplicates such a situation). Thus, a system with rare species almost certainly has high integrity. The converse is almost certainly not true; systems with high integrity may also lack rare species. So while the presence of rare species indicates a healthy ecosystem, their absence does not indicate distress. To explore this, we looked at the numbers of rare species in 401 quadrats representing wetlands from the Great Lakes through to the maritimes. Figure 5 (Moore *et al.* 1989) shows that rare species occurred exclusively in sites with biomass of less than 100 $g/0.25\ m^2$. Again, high biomass sites associated with enriched conditions had no recorded rare species.

Plant biomass

State variables which measure the integrated effects of many other processes may often be more useful than lower scale variables which measure the performance of a single species. For example, biomass would be an indicator of performance of the primary producers in general and may not only be relatively easy to monitor, but may have far more information than the biomass of a few selected species. In the case of wetlands, biomass appears to be an excellent indicator of trophic status. It may be a far better measure than measures of water chemistry since plants respond to the cumulative effects of nutrients available over a long period of time. This is not an original point since Clements (1935) pointed out that taking measuring instruments into the field may yield far less information than using the performance of organisms themselves as indicators.

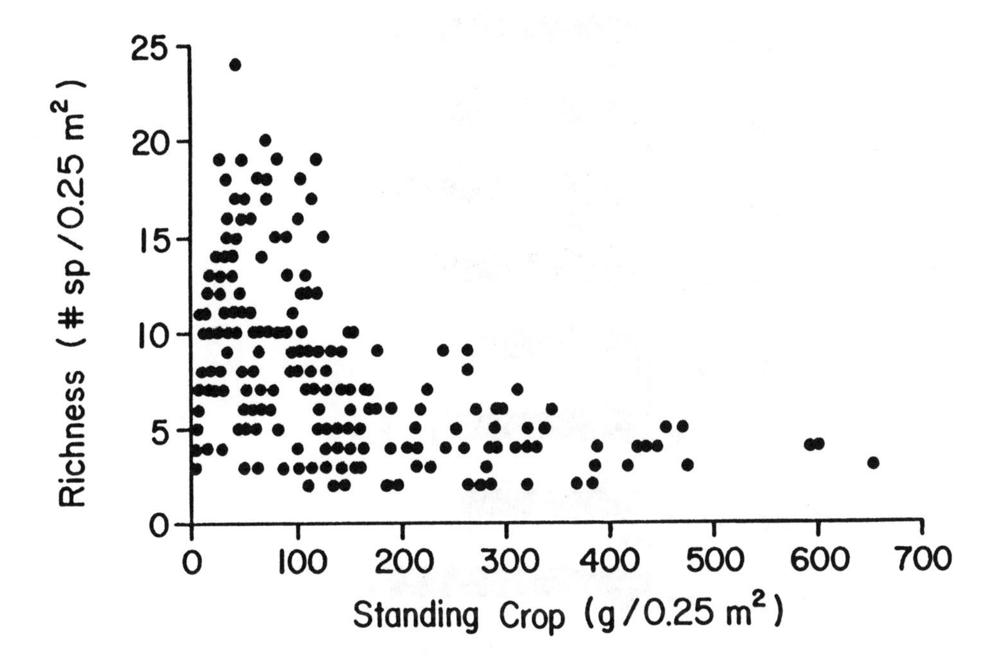

Figure 5: Number of nationally rare species versus standing crop in 401 0.25 m^2 quadrats from wetlands located in Ontario, Quebec and Nova Scotia (reprinted from Moore *et al.* 1989, with permission from Elsevier Applied Science Publishers).

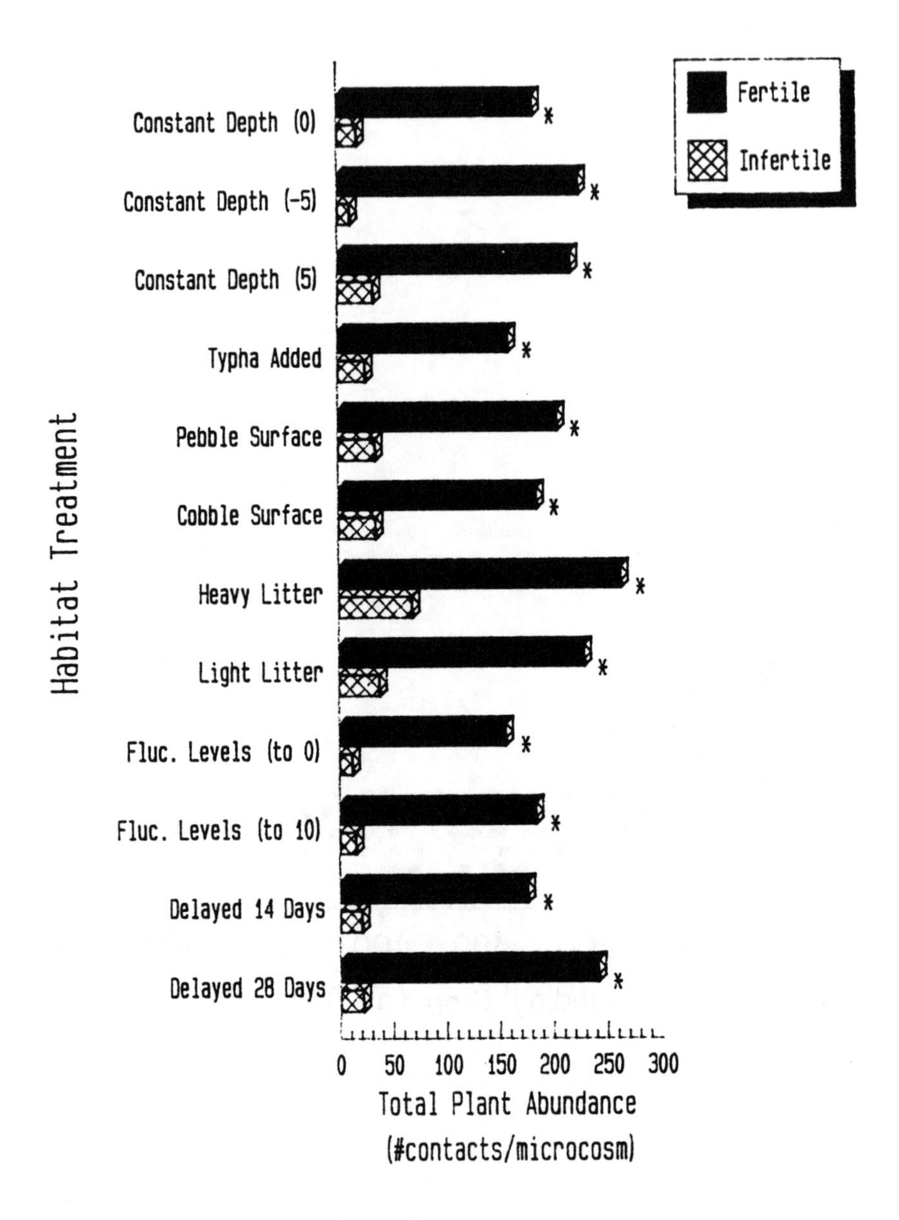

Figure 6: An indication of biomass (no. of plant contacts /mesocosm) after two growing seasons for twelve artificially created wetland habitats at two levels of fertility (After Moore and Keddy 1989).

Returning to the aforementioned mesocosm experiment (Moore 1990), Figure 6 shows the effects of enrichment in twelve different wetland habitat types. Enrichment increased biomass by a factor of 3 to 5 times.

Amphibian biomass

Components of an ecosystem which are known to sustain a large number of other components may also be useful indicators of ecosystem integrity. Because amphibians are an important food source for many higher trophic levels in wetlands, as well as being higher carnivores which are consumers of invertebrates, amphibian biomass may be an important indicator. Additionally, we know that amphibians have semipermeable skins and inhabit both the aquatic and terrestrial habitats, making them particularly sensitive to a wide range of contaminants. It has been noted that the populations of frogs and other amphibians are declining at a remarkable rate in North America, Central and South America, Europe, Asia, Africa and Australia (Blausstein and Wake 1990, Klassen 1991). Ongoing surveys in Alberta, for example, have shown that even the common leopard frog is rapidly disappearing. It is estimated that they have already disappeared from 85% of their former territory (Klassen 1991).

Conclusion

The above indicators are only a starting point for measuring the degree of integrity in wetlands. We wish to reemphasize that the critical limits of such indicators would be expected to vary among kinds of wetlands. Defining limits, and adding new indicators, are two research paths which can proceed in parallel.

Finally, we wish to argue for pragmatism (*sensu* Aune 1970) as a strategy in indicator development. Pragmatism requires us to make some preliminary decisions about indicators and critical limits as an essential first step to initiating the scientific process of using and refining them. The process of refinement would not only identify key areas for research, but would have immediate applications for ongoing conservation and restoration strategies.

ACKNOWLEDGEMENTS

We thank Stephen Woodley, Henry Regier and George Francis for the invitation to participate in this discussion. We also appreciated the helpful comments on the manuscript provided by Tom Nudds and Tom Whillans.

REFERENCES

Aune, B. 1970. *Rationalism, Empiricism and Pragmatism: An Introduction.* Random House.

Beanland, G.E. and P.N. Duinker. 1983. *An Ecological Framework for Environmental Impact Assessment in Canada.* Institute for Resource and Environmental Studies, Dalhousie University, and Federal environment assessment review office.

Blaustein, A.R., and D.B. Wake. 1990. Declining amphibian populations: a global phenomenon? *TREE.* 5:203-204.

Boston, H.L. and M.S. Adams. 1987. Productivity, growth and photosynthesis of two small 'isoetid' plants, *Littorella uniflora* and *Isoetes macrospora. Journal of Ecology.* 75:333-350.

Christie, W.J. 1974. Changes in the fish species composition of the Great Lakes. *Journal of the Fisheries Research Board of Canada.* 31:827-854.

Clements, F.E. 1935. Experimental ecology in the public service. *Ecology.* 16:342-363.

Ehrenfeld, J.G. 1983. The effects of changes in land-use on swamps of the New Jersey Pine Barrens. *Biological Conservation.* 25:353-375.

Environment Canada. 1987. The Canadian wetland classification system. National wetlands working group; Canada committee on ecological land classification (series no. 21). Lands conservation branch, Canadian Wildlife Service. EN 73-3/21E.

Environment Canada. 1991. A state of the environment report: A report on Canada's progress towards a national set of environmental indicators. EN 1-11/91-1E.

Evans, D.O., G.J. Warren and V.W. Cairns. 1990. Assessment and management of fish community health in the Great Lakes: synthesis and recommendations. *Journal of Great Lakes Research* . 16:639-669.

Hellawell, J.M. 1991. Development of a rationale for monitoring. in F.B. Goldsmith, editor. *Monitoring for Conservation and Ecology.* Chapman and Hall, London.

Holling, C.S. (ed.) 1978. *Adaptive Environmental Assessment and Management*. John Wiley and Sons.

Karr, J.R. 1987. Biological monitoring and environmental assessment: a conceptual framework. *Environmental Management*. 11:249-256.

Karr, J.R. 1991. Biological integrity: A long-neglected aspect of water resource management. *Ecological Applications*. 1:66-84.

Keddy, P.A. 1981. Vegetation with Atlantic coastal plain affinities in Axe Lake, near Georgian Bay, Ontario. *Canadian Field Naturalist*. 95:241-248.

Keddy, P.A. 1987. Beyond reductionism and scholasticism in plant community ecology. *Vegetation*. 69:209-211.

Keddy, P.A. 1989. *Competition*. Chapman and Hall, London.

Keddy, P.A. 1991. Biological monitoring and ecological prediction: from nature reserve management to national state of the environment indicators. In: F.B. Goldsmith, editor. *Monitoring for Conservation and Ecology*. Chapman and Hall, London.

Keddy, P.A. and A.A. Reznicek. 1986. Great Lakes vegetation dynamics: the role of fluctuating water levels and buried seeds. *Journal of Great Lakes Research*. 12:25-36.

Keddy, P.A. and I.C. Wisheu. 1989. Ecology, biogeography and conservation of coastal plain plants: some general principles from the study of Nova Scotia wetlands. *Rhodora*. 91:72-94.

Keddy, C.J., and T. McRae. 1989. Environmental databases for state of the environment reporting: Conservation and protection headquarters, Report No. 9, Strategies and scientific methods, SOE reporting branch, Canadian Wildlife Service, Environment Canada.

Keddy, P.A. 1990. The use of functional as opposed to phylogenetic systematics: A first step in predictive community ecology. In: *Biological Approaches and Evolutionary Trends in Plants*. Academic Press

Klassen, M. 1991. Losing their spots: No one knows why leopard frogs are disappearing from Alberta. *Nature Canada*. 20(1):9-11.

Lewontin, R.C. 1974. *The Genetic Basis of Evolutionary Change*. Columbia University, New York.

Lubchenco, J., A.M. Olson, L.B. Brubaker, S.R. Carpenter, M.M. Holland, S.P. Hubbell, S.A. Levin, J.A. MacMahon, P.A. Matson, J.M. Melillo, H.A. Mooney, C.H. Peterson, H.R. Pulliam, L.A. Real, P.J. Regal, and P.G. Risser. 1991. The sustainable biosphere initiative: An ecological research agenda. *Ecology*. 72:371-412.

Moore, D.R.J. 1990. Pattern and process in wetlands of varying standing crop: the importance of scale. Phd Thesis, University of Ottawa, Ottawa, Ontario.

Moore, D.R.J., and P.A. Keddy. 1989. Infertile wetlands: conservation priorities and management. In: M.J. Bardecki and N. Patterson, eds. *Wetlands: Inertia or Momentum?* Federation of Ontario Naturalists, Don Mills, Ontario.

Moore, D.R.J., P.A. Keddy, C.L. Gaudet, and I.C. Wisheu. 1989. Conservation of wetlands: do infertile wetlands deserve a higher priority? *Biological Conservation*. 47:203-217.

Patten, B.C. (ed.) 1990. Wetlands and shallow continental water bodies, Vol. 1: *Natural and Human Relationships*. SPB Academic Publishing by The Hague, The Netherlands.

Peters, R.H. 1991. *A Critique for Ecology*. Cambridge University Press, Cambridge.

Pielou, E.C. 1975. *Ecological diversity*. John Wiley and Sons, New York.

Rapport, D.J. 1989. Symptoms of pathology in the Gulf of Bothnia (Baltic Sea): ecosystem response to stress from human activity. *Biological Journal of the Linnean Society*. 37:33-49.

Rapport, D.J. 1990. What constitutes ecosystem health? *Perspectives in Biology and Medicine*. 33:121-132.

Rapport, D.J., C. Thorpe, and H.A. Regier. 1979. Ecosystem medicine. *Bulletin of the Ecological Society of America*. 60:180-182.

Rapport, D.J., J.A. Regier, and T.C. Hutchinson. 1985. Ecosystem behaviour under stress. *American Naturalist*. 125:617-640.

Rigler, F.H. 1982. Recognition of the possible: an advantage of empiricism in ecology. *Canadian Journal of Fisheries and Aquatic Sciences.* 39:1323-1331.

Ryder, R.A. 1990. Ecosystem health, a human perspective: definition, detection, and the Dichotomous Key. *Journal of Great Lakes Research.* 16:619-624.

Schaeffer, D.J., E.E. Herricks, and H.W. Kerster. 1988. Ecosystem health: Measuring ecosystem health. *Environmental Management.* 12:445-455.

Severinghaus, H.D. 1981. Guild theory development as a mechanism for assessing environmental impact. *Environmental Management.* 5:187-190

Sharitz, R.R. and J.W. Gibbons. (eds.) 1989. Freshwater wetlands and wildlife. U.S. Dept. of Energy; Office of Health and Environmental Research.

Tansley, A.G. 1914. Presidential address. *Journal of Ecology.* 2:194-203.

Terborgh, J., and S. Robinson. 1986. Guilds and their utility in ecology. In: J. Kikkawa and D.J. Anderson, eds. *Community Ecology: Pattern and Process.* Blackwell, Melbourne. pp. 65-90.

Thompson, D.Q., R.L. Stuckey, and E.B. Thompson. 1987. Spread, impact, and control of Purple Loosestrife (*Lythrum salicaria*). Fish and wildlife research 2, United States Department of the Interior, Fish and Wildlife Service. Washington, D.C.

Wisheu, I.C. and P.A. Keddy. 1989. Species richness - standing crop relationships along four lakeshore gradients: constraints on the general model. *Canadian Journal of Botany.* 67:1609-1617.

Wisheu, I.C., P.A. Keddy, D.R.J. Moore, S.J. McCanny, and C.L. Gaudet. 1992. Effects of eutrophication on wetland vegetation. In: The proceedings of the 'Wetlands of the Great Lakes' symposium, Niagara, New York, 1990.

Woodwell, G.M. and A.L. Rebuck. 1967. Effects of chronic gamma radiation on the structure and diversity of an oak-pine forest. *Ecological Monographs.* 37:53-69.

Applying the Concepts

Measuring Biological Integrity: Lessons from Streams

JAMES R. KARR
Institute for Environmental Studies, University of Washington
Seattle, Washington U.S.A.

The concepts of biological integrity and ecological health are not new. Over forty years ago, Aldo Leopold espoused a land ethic as he argued that "A thing is right when it tends to preserve the integrity, stability, and beauty of the biotic community. It is wrong when it tends to do otherwise" (Leopold 1949). Leopold's early hint at the pivotal role that integrity might play in environmental protection was passed over by two alternative perspectives: Pinchot's consumption oriented "resource conservation ethic" (the greatest good for the greatest number for the longest time) and Muir's "preservation ethic" (spiritual needs take precedence over material needs, Caldicott 1991). Over the past decade, the integrity concept has grown in influence, suggesting a transition in societal understanding of our relationship to the physical, chemical, and biological environment. That influence is likely to expand as our science and our societal value systems evolve.

This evolution can be seen in the changing perspective associated with the protection of water resources, especially increased attention to the protection of biological integrity of water resources (Karr 1991). Concrete examples of a rapidly evolving societal perspective include the development of National Program Guidance for Biological Criteria (USEPA 1990), the requirement for all states to have narrative biological criteria by 1993 (USEPA 1990), and Ohio's recent adoption of numerical criteria to improve the protection of biological resources (Ohio EPA 1990a). The clear statement of a ecological integrity objective in the National Parks Act in Canada and the adoption of an ecosystem objective in binational programs to protect the Laurentian Great Lakes are additional examples (Edwards *et al.* 1990). Yet another example can be seen in U. S. Environmental Protection Agency Administrator William Reilly's call for altering the emphasis in environmental protection from human health to reversing the destruction of soil, water, air, and biological resources. Caldicott (1991) refers to this as adoption of an ecological-evolutionary land ethic while I prefer an "ecological health" or "biological integrity ethic."

The Water Quality Act Amendments of 1972 (PL 92-500) first used the phrase biological integrity as it called for the restoration and maintenance of "the chemical, physical and biological integrity of the nation's waters." Three recent developments have increased attention to that objective: 1) recognition that water resources, particularly their biological components, continue to be degraded at alarming rates; 2) development of integrative biological approaches with considerable potential to recognize and reverse those trends; and 3) initiation of programs of biological monitoring by regulatory agencies. Although existing knowledge is not sufficient to make absolutely reliable resource decisions, neither can we afford the luxury of placing natural resource decisions on hold until that level of knowledge is available. I am confident that we have sufficient knowledge to improve the quality of resource decisions. The only question is do we have the will as a society to use it.

My charge for this paper is to outline the important lessons of the past decade relevant to measuring and protecting biological integrity or ecological health. I intend to address my comments to managers and to academics and I hope to avoid the stigma of being too esoteric to aid resource managers and of being too unsophisticated to reflect reality in complex natural systems.

The Role of Biology in Environmental Protection/Regulation

The ability to sustain a balanced biological community is one of the best indicators of the potential for beneficial use (defined broadly) of a water resource. However, biological systems and the influences of humans on those systems are complicated. No reasonable economist would suggest using the salary of university professors or government employees to assess the health of a nation's economy. Similarly, no single biological or chemical attribute is sufficient to evaluate ecological condition and complexity; consequently, we must seek a broadly based, ecologically sound multiparameter approach to understanding of biological integrity (Karr 1991).

Although this lesson seems obvious, some water resource planners continue to advocate narrow use of chemical criteria. Indeed, ecologists often debate the merit of systems versus population ecology or whether fish, invertebrates, or diatoms provide for the most rigorous resource evaluations. Such arguments strike me as equivalent to medical doctors arguing about whether to take the patient's temperature or measure their pulse rate. The most insightful assessments will come from a balance of all rigorous approaches and from the widest diversity of taxa (Karr 1991). Indeed, the more complex the water resource problem the more likely that a diversity of approaches will be needed to make informed decisions.

Perhaps the most important use of biological monitoring is in the establishment and implementation of numerical biocriteria; that is, quantitative expectations of the performance of biological communities. Use of numerical criteria in Ohio has enabled state agencies to refine aquatic life uses and criteria, conduct more accurate environmental assessments, broadened application of water quality standards to include non-chemical influences, and improved evaluation of difficult water resource issues (Ohio EPA 1990b). A number of other states have developed sophisticated approaches to the analysis of ecological health (Illinois EPA 1989, Michigan DNR 1990a, 1990b, Lyons 1992). Historically, water resource investigations have been plagued by high variability in water quality parameters. In use of biological monitoring, several states have found "virtually identical" metric scores for macroinvertebrate and fish biotic integrity analyses using different teams of investigators at the same sites (Michigan DNR 1990c). Habitat assessment scores are less consistent, a pattern found by Ohio EPA as well.

Most use of biological monitoring and biological criteria involves streams but recent efforts in lakes, reservoirs, wetlands and near coastal environments (Karr and Dionne 1991, Dionne and Karr in press, Thoma 1990, Marshall *et al.* 1987, Ryder and Edwards 1985, Edwards *et al.* 1990) suggest that the approach is more widely applicable. For example, Thoma (1990) found the highest biological integrity in Lake Erie estuaries with the fewest point sources and the least upstream agricultural use. Most importantly, use of biological integrity provides a balance to conventional chemical and physical evaluations while it focuses on the condition of the resource rather than some inadequate surrogate. The development of biological monitoring is driven in many areas by the need for rigorous and standardized evaluation of nonpoint source impacts and other circumstances where the upstream/downstream approach of point source evaluation is not appropriate.

The goal of integrative evaluations of water resource conditions is to protect ecological health or biotic integrity. Simply put, biological integrity is the ability of an environment to support and maintain a biota (both structural and functional performance) comparable to the natural habitats of the region (the 'harmonic community' of Edwards and Ryder 1990). The existence of that integrity suggests that "ecological health" is being protected. Ecological health is the condition when a system's inherent potential is realized, its condition is stable, its capacity for self repair, when perturbed, is preserved, and minimal external support for management is needed (Karr *et al.* 1986).

Evidence that biological integrity and ecological health are seriously threatened is widespread. Forty percent of the molluscs of the Ohio River drainage were listed as rare, endangered, or extinct by 1970 (Stansbery 1970). Two-thirds of the fishes of the Illinois River and 43% of Maumee River fish species have declined in abundance substantially or have disappeared in the last century (Karr *et al.* 1985). For the fishes of California, only 36% have stable populations, while 64% range from extinct (6%) to declining populations (22%, Moyle and Williams 1990). In North America, 364 fishes warrant protection because of their rarity (Williams *et al.* 1989). Despite massive expenditures to improve water quality, none of 251 fishes listed as rare in 1979 could be removed from the list in 1989 because of successful recovery efforts (Williams *et al.* 1989). In the U. S. Pacific Northwest, 214 native, naturally spawning Pacific salmon and steelhead stocks face "a high or moderate risk of extinction, or are of special concern" (Nehlsen *et al.* 1991).

As these examples demonstrate, water resources are still not adequately protected. Inadequately treated problems include toxics, nonpoint sources, habitat destruction, altered stream flows, and harvest and stocking practices. The limited use of biological factors in evaluating the quality of water resources perpetuates these problems and results in continuing declines in the health and the biological integrity of water resource systems.

The value of biological monitoring to assess the state of ecological health of a water resource system is well documented. Ohio EPA has been especially successful at incorporating numerical biocriteria into their water quality regulations (Yoder 1990). The most important findings of the Ohio efforts are as follows:

1. Biological criteria have a broad ability to assess and characterize a variety of chemical, physical, and biological impacts, as well as their cumulative impacts.
2. When biological criteria are integrated with chemical/toxicity assessments they can serve a broad range of environmental and regulatory programs including water quality standards (Table 1).
3. An integrated approach to surface water assessment provides increased environmental accuracy and is less likely to be underprotective of the water resource.
4. Impairment of Ohio waters is predominantly caused by nontoxic and nonchemical factors.
5. Numerical biological assessment approaches have greater precision and accuracy than narrative approaches.

Table 1. Regulation applications of biological criteria (from Dudley 1991).

Water Quality Standards and toxins control
National Pollution Discharge Elimination System (NPDES) permitting
Non-point source management and assessment
Natural resource damage assessment
Habitat protection
Attainment of aquatic life goals
Reporting trends (such as 305b reporting)
Stormwater and combined sewer overflow influence

Overall, the broader perspective of combining chemical/toxic assessment with biological monitoring is essential because controlling chemical water quality alone does not assure the integrity of water resources. Further, biological monitoring is useful in evaluating the ecological effects of toxics not just their toxicological effects.

The Constituents of Biological Integrity

Two weaknesses permeate most past approaches to protect water resources: 1) a narrow conception of the factors responsible for degradation and 2) a limited perspective on the components of biological integrity.

Narrow Conception of Degradation

Although several agencies have responsibility for protection of water resources, none (except perhaps Ohio EPA) have adopted a comprehensive approach that treats all variables that, when altered by human actions, result in the degradation of water resources. Five such classes of variables exist:

1. Water quality - temperature, turbidity, dissolved oxygen, organic and inorganic chemicals, heavy metals, toxic substances.
2. Habitat structure - substrate type, water depth and current velocity, spatial and temporal complexity of physical habitat.
3. Flow regime - water volume, temporal distribution of flows.
4. Energy source - type, amount, and particle size of organic material entering stream, seasonal pattern of energy availability.
5. Biotic interactions - competition, predation, disease, parasitism.

Any human activity that degrades one or more of those, degrades the quality of the water resource. Problems associated with water quality, the primary subject of efforts by USEPA and equivalent state agencies, include nutrients, organics, toxic compounds, turbidity, hardness, or altered temperature regimes which degrade the quality of a water resource.

Habitat structure is equally important but often ignored. Changes in the structure of the channel, its sinuosity, and the nature of the substrate, the amount of overhanging and instream cover, channel morphology, gradient, and riparian vegetation all play a major role in governing the physical structure of the stream channel and thus, influence the nature of the biological community in that system. The U. S. Fish and Wildlife Service (USFWS) and state fish and game agencies are concerned with habitat degradation.

All biological systems require a continuing supply of energy. That energy may derive from local photosynthesis or from organic materials flowing into the system (e.g., leaves falling into a stream from an adjacent forest during the autumn). Alteration in the nature of that energy source influences the quality of the water resource. For example, throughout this century society sought to reduce the supply of oxygen demanding waste added to water resource systems. However, organic material is critical to the biota of streams (input of leaves in autumn just mentioned). Because those biological systems have evolved to deal with a tremendous input of oxygen demanding wastes, they depend on those inputs. Where humans alter, through any of a variety of activities, the nature of the energy supplies coming into the stream, they may degrade the quality of the water resource. This issue is widely known in the theoretical ecology literature (Vannote *et al.* 1980, Minshall *et al.* 1983) and some have called attention to its importance in protecting water resource quality (Karr and Dudley 1981, Cummins *et al.* 1989); few agency programs devote attention to balancing energy sources.

Flow regime, including seasonal dynamics of flow and total annual flow, have a major influence on the quality of the water resource. The USFWS evaluates altered flow regimes, especially in western North America, with the instream-flow methodology.

Biotic interactions, including altered competition and predation, concludes the list of primary variables. Sport or commercial harvest of fish and introduction of exotics and/or hatchery reared fish degrade the quality of the water resource (Moyle 1989, Goodman 1991). Those activities are coming under increased scrutiny with expansion of the ecological-evolutionary land ethic (ecological health ethic). Karr *et al.* (1986, see also Karr 1991) provide a more detailed analysis of these five factors and how human actions impact the quality of water resources.

Overall, the determinants of water resource quality are complex, and the simplistic approach of making water cleaner is inadequate. The solution of water resource problems will not come from better regulation of chemicals or the development of better assessment tools to detect degradation caused by chemicals (Karr 1991).

Components of Biological Integrity

Two major aspects of biological systems — elements and processes — must be protected (Table 2). The most commonly cited attributes on the elements side are the species of plants and animals in aquatic communities. Additional critical elements include the genetic diversity within those species and the assemblages (communities, ecosystems, and landscapes) upon which those species depend. At the level of processes, a myriad of interactions ranging from energy flow and nutrient dynamics to evolution and speciation are critical to the maintenance of biotic integrity. Given sufficient technology, we could maintain species in zoos and genetic diversity in gene banks but in the absence of complex species assemblages and the processes that keep them in existence, we are not protecting biotic integrity. The advantages of this approach to protecting the quality of water resources are diverse (Table 3).

Recent attention to passage of biodiversity legislation, in my view, too narrowly focuses on the elements side of that ledger with the potential to perpetuate an error similar to that which developed under the Clean Water Act. That is, the focus on making water clean did not protect the resource from other human influences; production of crystal-clear, distilled water flowing down concrete channels does not insure a quality water resource. Biological integrity continues to decline under a solely chemical focus; a similar result is likely if the elements (species), but not the assemblages and processes of biotic integrity, are protected under proposed biodiversity legislation (Karr 1990).

An additional concept is critical to the success of efforts to protect biotic integrity. First, the attributes that reflect biotic integrity have values that vary geographically. Species richness, a component of the elements, varies geographically so the expectation of a high quality site must be varied. Standardization of chemical criteria (e.g., 5 ppm DO), although intuitively appealing, ignores the reality of geographical (and temporal) variation in the attributes of ecological systems. For example, total phosphorus standards should vary regionally and according to primary use among Minnesota lakes with values ranging from less than 15 to 90 ug/l (Heiskary *et al.* 1987). Thus, an important contribution of biological monitoring has been recognition of the need to set standards as a function of

Table 2. Components of biological integrity

ELEMENTS (sometimes referred to as Biological Diversity)

Genes
Populations/Species
Communities/Ecosystems
Landscapes

PROCESSES

Nutrient cycling
Photosynthesis
Water cycling
Speciation
Competition/Predation
Mutualisms

Table 3. Advantages of ambient biological monitoring (From Karr and Kerans 1991).

1. Broadly based ecologically

2. Provides biologically meaningful evaluation

3. Flexible for special needs

4. Sensitive to a broad range of degradation

5. Integrates cumulative impacts from point source, nonpoint source, flow alteration, and other diverse impacts of human society

6. Integrates and evaluates the full range of classes of impacts (water quality, habitat structure, etc.) on biotic systems

7. Direct evaluation of resource condition

8. Easy to relate to general public

9. Overcomes many weaknesses of individual parameter by parameter approaches

10. Can assess incremental degrees and types of degradation, not just above or below some threshold

11. Can be used to assess resource trends in space or time

local and regional expectations. Reference sites are selected to represent typical, least impacted areas within a homogeneous region. These relatively pristine areas are a benchmark against which a study site is evaluated.

Ambient biological monitoring offers unique opportunities to detect, analyze, and plan treatment of degraded resources. Use of biological monitoring grows in its sophistication and in the contexts in which it is used.

Assessing Biological Integrity

Numerous biological attributes can be measured to evaluate ecological health. Some are difficult and expensive to measure while others are not. Some provide reliable evaluations of biological conditions while others, perhaps because they are highly variable, are more difficult to interpret. Thus, the choice of attributes to be measured and assessed is critical to the success of any program to monitor biological conditions. Historically, most biological evaluations were designed to detect a narrow range of factors degrading water resources. The biotic index (Chutter 1972, Hilsenhoff 1987) is designed to detect the influence of oxygen demanding wastes ("organic pollution") or sedimentation, as is the saprobic index (Kolkwitz and Marrson 1907) developed early in this century.

With increased understanding of the complexity of biological systems and complex influences of human actions on those systems, more integrative approaches to assessing biological integrity have developed. Some (Ulanowicz 1990, Kay 1990, Kay and Schneider in press) advocate use of thermodynamics while others concentrate on richness or diversity (Wilhm and Dorris 1968). Some have argued that the thermodynamic approach is more sophisticated but I am not convinced. Perhaps they equate sophistication with mathematical complexity. I prefer an integrative approach that combines assessment of the extent to which either the elements or the processes of biological integrity have been altered; that is, the type of scientific sophistication that is appropriate to the protection of biological integrity should represent a broad diversity of ecological attributes not just thermodynamics or species richness.

Two such approaches are currently available: the Index of Biotic Integrity (IBI) developed in 1981 (Karr 1981) and expanded and modified for use in many areas (Ohio EPA 1988, Plafkin *et al.* 1989) and the dichotomous key indicator species approach developed for the Laurentian Great Lakes (Ryder and Edwards 1985, Edwards and Ryder 1990). Both base a judgement of ecological health on a range of attributes of biotic communities. (The lake trout as an indicator species is a bit of a misnomer because the dichotomous key developed for assessment of the Lake Superior ecosystem is much broader than an analysis of the single trout species.)

Perhaps the most striking difference in use of these two approaches is the spatial scale that they consider. The IBI can be used for local or regional areas while the dichotomous key focuses on the entire lake. Recent efforts to develop a similar key for Lake Ontario apparently explore a mixed approach to deal with both cool and warm water components of the lake (Edwards and Ryder 1990). Scaling for evaluation of local conditions within the lake (e.g., in single estuaries) has not been attempted. In addition, the cost of obtaining the required data may be high and impractical except in the context of a large sport or commercial resource. Its use of ecological principles and understanding of the study system is excellent.

Because I am more familiar with the foundations of the Index of Biotic Integrity (IBI), I turn now to a brief review of its ecological underpinnings. The IBI approach builds on the work of many scientists going as far back as Forbes and Richardson (1919, Richardson 1928). They, for example, demonstrated that biological assessments can be both direct and accurate in assessing environmental conditions. They also concluded that numerical abundances are not as useful as relative abundances in assessing biological conditions, and the occurrence of specific taxa was especially useful. The IBI methodology and its conceptual clones have been used throughout North America (Miller *et al.* 1988) and in Europe (Oberdorff and Hughes 1992), including efforts to use it with other taxa (e.g., invertebrates, Ohio EPA 1988, Plafkin *et al.* 1989, Karr and Kerans 1991, Kerans and Karr in review, Kerans *et al.* in press).

Most efforts to develop an IBI approach include suites of metrics that represent three major classes of biological attributes. Species richness and composition metrics evaluate the extent to which some of the expected elements of biotic integrity are present while several trophic composition metrics evaluate some of the process attributes, notably those associated with functional (food chain) conditions. The third set of metrics involves overall population sizes and the condition (health) of individuals.

The suite of IBI metrics has several other important attributes. Each metric is an expression of a known influence of human activities on the characteristics of the resident biota. Examples include the observation that, with increased human disturbance, total species richness, number of intolerant species, and number of trophic specialists decline, while the number of trophic generalists increases. Detailed studies of patterns in nature converts these from hypotheses to assumptions.

The complexity of ecological systems limits the likelihood that any single attribute can be used to assess all forms of degradation and be sensitive across the full range of degradation. Thus, a suite of metrics is more likely to provide insight about the conditions at a site than is any single attribute. Further, by assessing a number of attributes the overall index provides a number of more or less independent evaluations of site quality (Figure 1). The addition of a number of metrics improves the resolving power of the individual metrics, including precision over a wider range of integrity conditions. As a result, stronger inferences can be made when a number of metrics are used. No single metric can assess the extent of degradation with absolute certainty. A combination of metrics, each with an accuracy of perhaps 70-80%, narrows the range and improves the precision of the estimates of biological integrity at a site.

Some have argued that redundancy among the metrics is to be avoided. Rather, I suggest that it is appropriate to mimic the normal redundancy of natural systems. That redundancy is especially useful when different metrics have sensitivities at different positions along the overall biotic integrity gradient (e.g., total number of species, number of intolerant species, and percent green sunfish - Figure 2). An additional possibility is to develop information on the association between degradation of specific metrics or sets of metrics and certain human influences. For example, metrics a, d, and g might show degradation in the presence of toxic contamination while metrics b, c, d, and i are indicative of habitat degradation. Knowledge of suites of metrics with predictable associations with specific causes of degradation, termed "degradation signatures," could aid in diagnosing and resolving water resource problems.

Evaluation of a number of attributes of invertebrates communities to screen for potential metrics in an invertebrate IBI is also underway (Karr and Kerans 1991, Kerans *et al.* in press). Earlier developments of invertebrate community indices have, in my view, failed to exploit the full range of information available in samples of invertebrates. For example, the range of ecological attributes included in most invertebrate approaches is limited when compared to the fish IBI (Karr 1991). Ecological classifications such as trophic category, functional group, feeding mechanism, and habit of benthic invertebrates show considerable promise (Karr and Kerans 1991) in our analysis of invertebrate data collected in tributaries of the Tennessee River.

We have found that the total number of invertebrate taxa declines with human influence and the number of collector/filterer organisms tends to increase in degraded sites (Figure 3). The number of the mollusc, *Corbicula,* increased rather strikingly in the most degraded sites. We do not consider these early correlations to be definitive. Rather, they are indicative of relationships that deserve more careful scrutiny for possible use in a more general invertebrate IBI. We now seek other data sets to

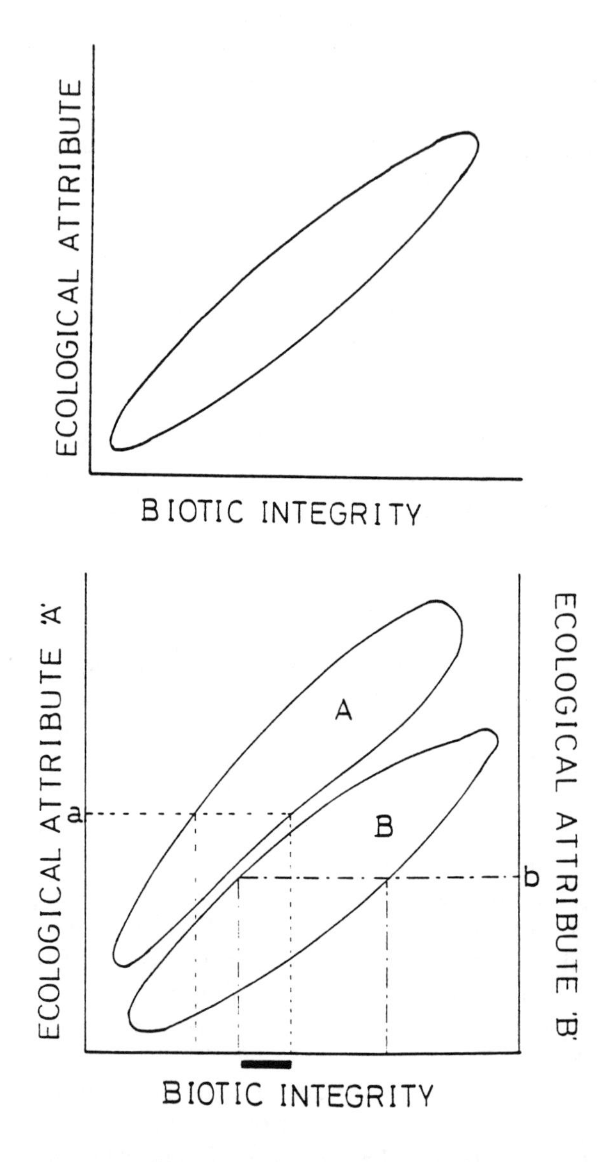

Figure 1: Conceptual depiction of the relationship between a single ecological attribute (IBI metric-upper panel) and two ecological attributes (lower panel) and biotic integrity. Note that simultaneous use of two metrics narrows the identified biotic integrity level (dark horizontal bar) relative to use of a single attribute (metric).

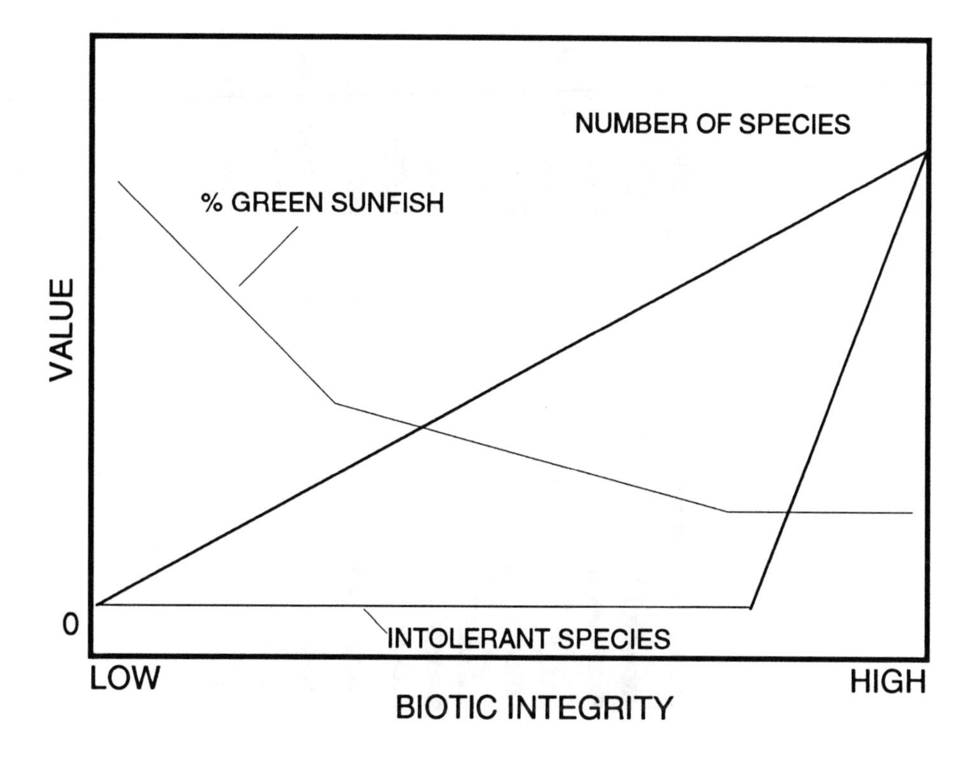

Figure 2: Conceptual depiction of the range of sensitivity of three IBI metrics across the gradient from low to high biotic integrity.

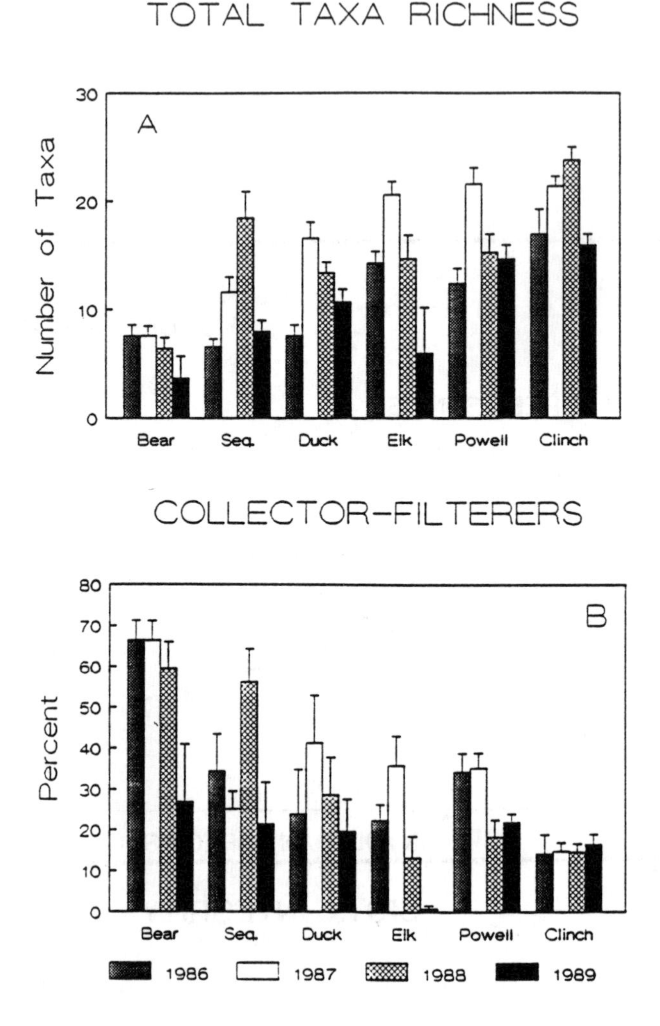

Figure 3: Mean and standard error bars for two potential invertebrate IBI metrics for six Tennessee River tributaries during four consecutive years (1986-1989). Note that the rivers are plotted from low quality (Bear Creek) to highest quality (Clinch River). A. Total Taxa Richness. B. Collector-Filters. Note that one decreases (A) while the other increases (B) with degradation.

determine if the pattern in a few Tennessee River tributaries can be generalized more widely. We started with 28 hypotheses (potential metrics), and have selected 13 for inclusion in a benthic index of biotic integrity (Kerans and Karr in review).

Principles of Biological Monitoring

A sound biological monitoring program designed to assess biological integrity should have several attributes. A firm conceptual foundation broadly based in ecological principles is essential, as is a multidimensional assessment (individual, population, species composition, trophic structure) that incorporates both the elements and processes of biotic integrity. The use of the concept of reference condition, a condition against which a site is evaluated, is also important. Levels of quality control and quality assurance must be developed as in any analysis that may be used to protect environmental health. The development of standard sampling protocols is essential to the comparison of data in space and time. Finally, biological monitoring must go beyond the collection and tabulation of high quality data to the creative analysis and synthesis of relevant biological attributes. For too many years, biological data have been collected and stored in file cabinets without adequate exploration and synthesis of patterns in those data. To do less is to reduce the potential value and impacts of biological data and, perhaps worse, to relegate the resource to continuing degradation.

The time is ripe for implementing new programs to protect biotic integrity. Several recent innovations are responsible for this opportunity. The evolution of a broader set of societal goals, the use of a broader range of scientific disciplines, and longer term planning perspectives are examples of recent innovations. Further, the tendency to develop environmental protection programs that are implemented in a wider geographic context fosters the transition to the use of an ecological health or biotic integrity ethic.

These opportunities also present a number of challenges (Karr in press). Considerable effort must also be directed to the documentation of, and interpretation of natural variation. Unfortunately, a tendency has developed to consider the mean in biological data as the signal and variation as noise. I submit that in many cases, the nature of the variation may also be a part of the signal. For example, a number of researchers have shown that variation increases in situations with major human impacts (Karr *et al.* 1987, Steedman 1988, Rankin and Yoder 1990), suggesting that the natural buffering capacity of biological systems is reduced by anthropogenic influences. Another challenge is the need to develop biological assessment approaches that increase the likelihood that degradation will be detected, independent of the causes of that decline. As the influence of human society expands, a growing proportion of environmental degradation derives from a complex of cumulative impacts. Effective evaluation of hypotheses about human influences is critical to the development of a robust index of biotic integrity. Biologists must become primary contributors to this dialogue (Karr 1991). They must overcome an inherent skepticism about the simplicity of biological assessment while they also guard against the cookbook application of biotic monitoring by individuals unfamiliar with ecological processes.

Summary

Growing concern for environmental degradation indicates a move to a "biological integrity" or "ecological health ethic" to replace Pinchot's resource conservation ethic and Muir's preservation ethic. This shift is especially obvious in the water resource arena where early efforts concentrated on a few contaminants and later expanded to treatment of a growing array of toxic materials. Chemical and toxic evaluations have failed to halt the degradation in the quality of water resources. The analogy of the instability of a two legged stool comes to mind. The addition of a third leg, direct monitoring of the biological assemblages of the water resource, is likely to create a more stable foundation for the protection of the resource. A tripod is a more appropriate analogy because the relative lengths of the legs can be varied to fit the local "landscape" or configuration of water resource problems.

REFERENCES

Caldicott, J. B. 1991. Conservation ethics and fishery management. *Fisheries*. (Bethesda) 16(2):22-28.

Chutter, F. M. 1972. An empirical biotic index of the quality of water in South African streams and rivers. *Water Research* . 6:19-20.

Cummins, K. W., M. A. Wilzbach, D. M. Gates, J. B. Perry, and W. B. Taliaferro. 1989. Shredders and riparian vegetation. *BioScience*. 39(1):24-30.

Dionne, M. and J. R. Karr. In press. Ecological monitoring of fish assemblages in Tennessee River reservoirs. In: D.H. McKenzie, D.E. Hyatt, and V.J. McDonald (eds.). *Ecological Indicators*. Elsevier Science Publishers, Essex, England, pp. 259-281.

Dudley, D. R. 1991. A state perspective on biological criteria in regulation. In: *Biological Criteria: Research and Regulation*. 1991. U. S. Environmental Protection Agency, Washington, D.C., EPA - 440/5-91-005, pp. 15-18.

Edwards, C. J. and R. A. Ryder. 1990. Biological surrogates of mesotrophic ecosystem health in the Laurentian Great Lakes. Report to the Great Lakes Science Advisory Board, International Joint Commission, Windsor, Ontario.

Edwards, C. J., R. A. Ryder, and T. R. Marshall. 1990. Using lake trout as a surrogate of ecosystem health for oligotrophic waters of the Great Lakes. *Journal Great Lakes Research* . 16:591-608.

Forbes, S. A. and R. E. Richardson. 1919. Some recent changes in Illinois River biology. *Bulletin Illinois Natural History Survey*. 13:139-156.

Goodman, B. 1991. Keeping anglers happy has a price. *BioScience*. 41:294-299.

Heiskary, S. A., C. B. Wilson, and D. P. Larsen. 1987. Analysis of regional patterns in lake water quality: using ecoregions for lake management in Minnesota. *Lake Reservoir Management*. 3:337-344.

Hilsenhoff, W. L. 1987. An improved biotic index of organic stream pollution. *Great Lakes Entomologist*. 20:31-39.

Illinois Environmental Protection Agency. 1989. Biological stream characterization (BSC): a biological assessment of Illinois stream quality. *Special Report #13*. Illinois State Water Plan Task Force. Illinois Environmental Protection Agency, Springfield, IL.

Karr, J. R. 1981. Assessment of biotic integrity using fish communities. *Fisheries (Bethesda)*. (6):21-27.

Karr, J. R. 1990. Biological integrity and the goal of environmental legislation: lessons for conservation biology. *Conservation Biology*. 4:244-250.

Karr, J. R. 1991. Biotic integrity: A long-neglected aspect of water resource management. *Ecological Applications*. 1:66-84.

Karr, J. R. In press. Biological monitoring: challenges for the future. In: S. Loeb and A. Spacie (eds.). *Biological Monitoring of Freshwater Ecosystems*. Lewis Publ. Co., Ann Arbor, MI.

Karr, J. R. and M. Dionne. 1991. Designing surveys to assess biological integrity in lakes and reservoirs. In: *Biological Criteria: Research and Regulation*. U. S. Environmental Protection Agency, Washington, D.C., EPA - 440/5-91-005, pp. 62-72.

Karr, J. R. and D. R. Dudley. 1981. Ecological perspective on water quality goals. *Environmental Management*. 5:55-68.

Karr, J. R. and B. L. Kerans. 1991. Components of biological integrity: their definition and use in development of an invertebrate IBI. In: W. S. Davis and T.P. Simon (eds.). *Proceedings Midwest Pollution Control Biologists Meeting*. U. S. Environmental Protection Agency, Chicago, IL, EPA - 905/R-92/003, pp. 1-16.

Karr, J. R., L. A. Toth, and D. R. Dudley. 1985. Fish communities of midwestern rivers: a history of degradation. *BioScience*. 35:90-95.

Karr, J. R., K. D. Fausch, P. L. Angermeier, P. R. Yant, and I. J. Schlosser. 1986. Assessment of biological integrity in running water: a method and its rationale. Illinois Natural History Survey Special Publication No. 5, Champaign, IL. 28pp.

Karr, J. R., P. R. Yant, K. D. Fausch, and I. J. Schlosser. 1987. Spatial and temporal variability of the index of biotic integrity in three midwestern streams. *Transactions American Fisheries Society*. 116:1-11.

Kay, J. J. 1990. A non-equilibrium thermodynamic framework for discussing ecosystem integrity, pp. 209-237. In: C. J. Edwards and H. A. Regier (eds.). An ecosystem approach to the integrity of the Great Lakes in turbulent times. *Special Publication #90-4*, Great Lakes Fishery Commission, Ann Arbor, MI.

Kay, J. J. and E. D. Schneider. In press. Thermodynamics and measures of ecological integrity. In: D.H. McKenzie, D.E. Hyatt, and V.J. McDonald (eds.). *Ecological Indicators*. Elsevier Science Publishers, Essex, England.

Kerans, B.L. and J.R. Karr. In review. An evaluation of invertebrate attributes and a benthic index of biotic integrity for Tennessee Valley Rivers. Submitted to *Journal North American Benthological Society*.

Kerans, B. L., J. R. Karr. In press. Aquatic invertebrate assemblages: Spatial and temporal differences among sampling protocols. *Journal North American Benthological Society*.

Kolkwitz, R. and M. Marsson. 1908. Ökologie der pflanzlichen saprobien. *Berichte der Deutschen Botanischen Gesellschatt*. 26:505-519.

Leopold, A. 1949. *A Sand County Almanac: and Sketches Here and There*. Oxford University Press, New York.

Lyons, J. 1992. Using the index of biotic integrity (IBI) to measure environmental quality in warm water streams of Wisconsin. U.S. Department of Agriculture Forest Service, North Central Forest Experiment Station, St. Paul, MN. General Technical Report NC-149.

Marshall, T. R., R. A. Ryder, C. J. Edwards, and G. R. Spangler. 1987. Using the lake trout as an indicator of ecosystem health - application of the dichotomous key. *Technical Report #49*. Great Lakes Fishery Commission, Ann Arbor, MI.

Michigan Department of Natural resources. 1990a. Qualitative biological and habitat survey protocols for wadable streams and rivers. *GLEAS Procedure #51*. Michigan Department of Natural Resources, Great Lakes and Environmental Assessment Section, Surface Water Quality Division, East Lansing, MI.

Michigan Department of Natural Resources. 1990b. Reference site scores for wadable streams, 1990. MI/DNR/SWQ-90/119. Michigan Department of Natural Resources, East Lansing, MI.

Michigan Department of Natural Resources. 1990c. A quality assurance evaluation of the qualitative biological and habitat survey protocols for wadable streams and rivers. MI/DNR/SWQ-90/118. Michigan Department of Natural Resources, East Lansing, MI.

Miller, D. L., P. M. Leonard, R. M. Hughes, J. R. Karr, P. B. Moyle, L. H. Schrader, B. A. Thompson, R. A. Daniels, K. D. Fausch, G. A. Fitzhugh, J. R. Gammon, D. B. Haliwell, P. A. Angermeier, and D. J.

Orth. 1988. Regional applications of an index of biotic integrity for use in water resource management. *Fisheries (Bethesda).* 13(5):12-20.

Minshall, G. W., R. C. Peterson, K. W. Cummins, T. L. Bott, J. R. Sedell, C. E. Cushing, and R. L. Vannote. 1983. Interbiome comparison of stream ecosystem dynamics. *Ecological Monographs.* 51:1-25.

Moyle, P. B. 1989. Message from the president. Newsletter of the Introduced Fish Section, *American Fisheries Society.* 9(2):1-2.

Moyle, P. B. and J. E. Williams. 1990. Biodiversity loss in the temperate zone: decline of the native fish fauna of California. *Conservation Biology.* 4:275-284.

Nehlsen, W., J. E. Williams, and J. A. Lichatowich. 1991. Pacific salmon at the crossroads: stocks at risk from California, Oregon, Idaho, and Washington. *Fisheries (Bethesda).* 16(2):4-21.

Oberdorff, T. and R. M. Hughes. 1992. Modifications of an index of biotic integrity based on fish assemblages to characterize rivers of the Seine Normandie Basin, France. *Hydrobiologie.* 228:117-130.

Ohio Environmental Protection Agency. 1988. Biological criteria for the protection of aquatic life. Ohio Environmental Protection Agency, Division of Water Quality Monitoring and Assessment, Surface Water Section, Columbus, OH.

Ohio Environmental Protection Agency. 1990a. *Ohio Water Quality Standards.* Chapter 3745-1 of the Administrative code. Ohio Environmental Protection Agency, Columbus, OH.

Ohio Environmental Protection Agency. 1990b. The use of biocriteria in the Ohio EPA surface water monitoring and assessment program. Ohio Environmental Protection Agency, Columbus, OH.

Plafkin, J. L., M. T. Barbour, K. D. Porter, S. K. Gross, and R. M. Hughes. 1989. Rapid bioassessment protocols for use in streams and rivers: benthic macroinvertebrates and fish. U. S. Environmental Protection Agency, Washington, DC EPA/444/4-89-001.

Rankin, E. T. and C. O. Yoder. 1990. The nature of sampling variability in the index of biotic integrity (IBI) in Ohio streams. Pp. 9-18 in W. S. Davis (ed.). *Proceedings of the 1990 Midwest Pollution Control Biologists*

Meeting. U. S. Environmental Protection Agency, Chicago, IL. EPA 905/9-90-005.

Richardson, R. E. 1928. The bottom fauna of the Middle Illinois River, 1913-1925: its distribution, abundance, valuation, and index value in the study of stream pollution. *Bulletin Illinois Natural History Survey*. 17:387-472.

Ryder, R. A. and C. J. Edwards (eds.). 1985. A conceptual approach for the application of biological indicators of ecosystem quality in the Great Lakes basin. *Report to the Great Lakes Science Advisory Board*. International Joint Commission and Great Lakes Fishery Commission, Windsor, Ontario.

Stansbery, D. H. 1970. Eastern freshwater mollusks (I) the Mississippi and St. Lawrence River systems. *Malacologia*. 10:9-22.

Steedman, R. J. 1988. Modification and assessment of an index of biotic integrity to quantify stream quality in Southern Ontario. *Canadian Journal Fisheries Aquatic Sciences*. 45:492-501.

Thoma, R. F. 1990. A preliminary assessment of Ohio's Lake Erie estuarine fish communities. Ohio Environmental Protection Agency, Columbus, OH.

Ulanowicz, R. E. 1990. Ecosystem integrity and network theory. Pp. 69-77 in C. J. Edwards and H. A. Regier (eds.). An ecosystem approach to the integrity of the Great Lakes in Turbulent Times. *Special Publication #90-4*. Great Lakes Fishery Commission, Ann Arbor, MI.

United States Environmental Protection Agency. 1990. Biological criteria: national program guidance for surface waters. United States Environmental Protection Agency, Washington, DC. EPA-440/5-90-004.

Vannote, R. L., G. W. Minshall, K. W. Cummins, J. R. Sedell, and C. E. Cushing. 1980. The river continuum concept. *Canadian Journal Fisheries Aquatic Sciences*. 37:130-137.

Wilhm, J. L. and T. C. Dorris. 1968. Biological parameters of water quality criteria. *BioScience*. 18:477-481.

Williams, J. E., J. E. Johnson, D. A. Hendrickson, S. Contreras-Balderas, J. D. Williams, M. Navarro-Mendoza, D. E. McAllister, and J. E. Deacon. 1989. Fishes of North America: endangered, threatened, or of special concern: 1989. *Fisheries (Bethesda).* 14(6):2-20.

Yoder, C. O. 1990. Some questions and concerns about biological criteria based on experiences in Ohio. Ohio Environmental Protection Agency, Columbus, OH.

Monitoring for Ecosystem Integrity

R.E. MUNN
Institute for Environmental Studies, University of Toronto
Toronto, Ontario

Introduction

Ecological integrity, the ecosystem approach, and ecologically sustainable development are phrases that have been increasingly used in recent years. These terms have two features in common: they relate to long term ecosystem management, and the ecosystems involved generally include humans. However, the relevant literature suggests that none of the phrases can easily be applied operationally; in particular, the first and the third phrases are in the nature of aspirations or social goals. If we achieved a state of ecological integrity or sustainable development, we probably wouldn't recognize it at the time.

Of the key words mentioned above, development is the most difficult one to define, at least in the environment-development context. Many people think of development as growth in the sense of an increase in quantity/yield (with prosperity for all!) but because such growth cannot continue indefinitely, sustainable development is then a contradiction. A way out of this dilemma is to define development as evolution (Munn 1990, IUCN 1991), which is a perfectly good synonym. In the ecological context, evolution has been taking place over millennia, with very considerable success in most cases, and development can then be said to have been sustained. (Growth can also mean an increase in quality but this involves a value judgement, and evolution does not necessarily lead to the development of more valuable ecosystems — just to ones that have adapted successfully to changes in their environment.)

Ecosystems that have evolved successfully are said to have integrity. Although the phrase ecosystem integrity is not well defined, there seems to be a consensus on the properties that such a system should exhibit (Steedman and Regier 1990):

- strong, energetic, natural ecosystem processes and not severely constrained;
- self-organizing in an emerging, evolving way;
- self-defending against invasions by exotic organisms;

- biotic capabilities in reserve to survive and recover from occasional severe crises;
- attractiveness, at least to informed humans;
- productive of goods and opportunities valued by humans.

As pointed out by Kay (1991), ecosystems are not static, and thus "it is not possible to identify a single organizational state of the system which corresponds to integrity." So we have a moving target, in which ecosystems seek equilibrium in an environment that is continually changing. A particular problem facing society in the 1990s is that externalities are beginning to change more rapidly than ever before, and unprecedented transformations are expected in the 21st century. Can the evolutionary processes keep pace? Note that it is these processes that need to be sustained, not the current ecosystem state.

Two other ideas should be mentioned. The first is that of resilience (Holling 1973), which is a measure of the ability of an ecosystem to absorb shocks and to maintain its integrity even if the shocks are so great that the system shifts into a new mode. However, there is little guidance on how to measure resilience. Thus, it is not possible to predict the threshold level of stress that will cause a particular ecosystem to shift into a new mode (Regier 1989).

The second idea is the need to adopt an ecosystem approach toward long term environmental management (Edwards and Regier 1990). In the context of the Great Lakes Basin, this phrase has come to mean (IJC 1989, Harris *et al.* 1990):

- an interdisciplinary approach,
- a systemic perspective of the natural and cultural ecosystems that are intertwined in the Basin, and
- adaptive management that seeks to achieve cultural and natural integrity of the Basin.

Obviously, if management goals are to preserve ecosystem integrity and to promote sustainable development, an ecosystem approach is essential. Piecemeal management of a river basin, for example, leads to conflicts in resource use that can only be damaging to the long term integrity of the system.

If this conceptual framework is accepted, it follows that pathways leading towards ecologically sustainable development require that ecological integrity be preserved and that ecological resilience be maintained; this can only be achieved if an ecosystem approach is adopted. Thus, the various phrases are very closely linked.

Having dealt with these conceptual issues, admittedly in a reductive way, we can turn to the main subject of the chapter — monitoring for ecosystem integrity. But it must be admitted at the outset that an appropriate monitoring strategy is difficult to formulate except in the special case when externalities are constant or changing very slowly. Unfortunately, the appropriate time horizon for assessing ecosystem integrity is half a century, which makes the steady state assumption dubious.

Criteria for Designing Systems for Monitoring Ecosystem Integrity

A number of desirable properties of ecosystem integrity indicators can be mentioned:

1. ***Valued ecosystem components*** *(Beanlands and Duinker, 1983):* Not everything that could conceivably relate to ecosystem integrity can be monitored. So it is necessary to be selective, recognizing however that the selection process should involve the principal stakeholders in the long term future of the region. Here it should be emphasized that there is no single pathway leading towards ecosystem integrity and sustainable development and that societal goals often change from generation to generation. In addition, it is important to note that because each stakeholder will be interested in different ecosystem components and at different levels of complexity, some very considerable practical difficulties will have to be overcome in the design of a monitoring system that will satisfy the range of user needs.
2. ***Conceptual frameworks:*** At least some general conceptual frameworks of ecosystem behaviour exist, and the selected indicators of integrity should reflect current understanding of these ideas. For example, stress-response ecosystem models and actor system dynamics will be helpful (Rapport 1983, Francis and Regier 1991).
3. ***Externalities:*** Threats to ecosystem integrity can come from actions within a region ("soiling one's own nest," e.g., poor city planning, ground water pollution) or externally (global climate change, shifts in world markets, etc.). It is, in fact, contrary to ecological principles to treat a region, even one as large as the Great Lakes Basin, as a closed system. Thus, the ecosystem approach must include externalities, some of which will need to be monitored.

   In this same vein, sectoral approaches have their drawbacks. In the first place, ecosystem boundaries, e.g., of forest and cropland, do not often correspond with watershed boundaries. Secondly, the various economic sectors strongly interact with one another, and plans for long term sustainable development of the forest sector, for example, must take other human activities into account.

These ideas have been elaborated by Allen *et al.* (1987) and Allen and Hoekstra (1990).

4. ***Early warning capability:*** One of the most important properties of an indicator of ecosystem integrity is that it provide early warning of trouble ahead. Because of naturally occurring variability in the environment, managers often have difficulty in deciding whether relatively rare events (e.g., droughts, smog episodes or fish kills), which may have recently occurred more frequently than usual, herald a long term shift in baseline conditions. Historical time series of environmental indicators, e.g., rainfall, oxidants and fish populations, are uncertain predictors of future events, but if a manager waits until the statisticians can provide 95% confidence of a trend, the possibility of implementing an effective management strategy no longer exists. Thus, it is important to have at least a conceptual model of the processes involved, and to monitor as far back in the cause-effect chain as possible. For example, severe summer smog episodes are due to a combination of rare meteorological conditions (heat waves, clear skies, light winds) and increased emissions of oxides of nitrogen from power stations, which are operating at peak loading to meet the energy demands of air conditioning equipment. Thus, an appropriate early warning indicator of an increase in smog episodes is an upward trend in the number of air conditioners (automobile, domestic, commercial) in the region. Long term time series of ozone concentrations will not provide sufficient warning to permit the development of effective policies to prevent the worsening of summer smog. (A similar type of early warning indicator in the water pollution field might be the number of flush toilets within a watershed.)
5. ***Reproducibility:*** The measurements should be objectively obtained, and should be perceived to be so by the various stakeholders interested in the long term sustainability of the region.

Ecosystem Integrity in the Context of the UNESCO World Heritage Convention

The guidelines for implementation of the UNESCO World Heritage Convention (UNESCO 1988) refer in several places to the need to preserve the integrity of cultural and natural heritages. Specifically in paragraph 36(b), the following "conditions of integrity" are laid out:

(i) The sites should contain all or most of the key interrelated and interdependent elements in their natural relationships; for example, an "ice age" area would be expected to include the snow field, the glacier itself and samples of cutting patterns, deposition and colonization (striations, moraines, pioneer stages of plant succession, etc.).

(ii) The sites should have sufficient size and contain the necessary elements to demonstrate the key aspects of the process and to be self-perpetuating. For example, an area of tropical rain forest may be expected to include some variation in elevation above sea level, changes in topography and soil types, river banks or oxbow lakes, to demonstrate the diversity and complexity of the system.

(iii) The sites should contain those ecosystem components required for the continuity of the species or of the other natural elements or processes to be conserved. This will vary according to individual cases; for example, the protected area of a waterfall would include all, or as much as possible, of the supporting catchment area; or a coral reef area would include the zone necessary to include siltation or pollution through the streamflow or ocean currents which provide its nutrients.

(iv) The area containing threatened species should be of sufficient size and contain necessary habitat requirements for the survival of the species.

(v) In the case of migratory species, seasonal sites necessary for their survival, wherever they are located, should be adequately protected.

(vi) The sites should have adequate long term legislative, regulatory or institutional protection. They may coincide with or constitute part of existing or proposed protected areas such as national parks.

(A few nonessential words/sentences have been omitted from this quotation.)

In this example, specific indicators of ecosystem integrity are implied by the guidelines, e.g., the presence of key interrelated elements, sufficient size, existence of the ecosystem components required to ensure continuity of species. An appropriate strategy would thus be to monitor regularly (2-yearly intervals?) to determine trends, particularly the extent and impact of biological (including human) intrusions. Of course, the UNESCO criteria for the integrity of heritage sites might not be complete. Refinement of the criteria through continuing research and monitoring should be encouraged.

Monitoring of Regional Ecosystem Integrity

The UNESCO World Heritage Convention applies to rather small scale "protected" areas. The more important case is the large "patchwork quilt" region, which is, of course, subject not only to normal environmental controls but also to strong economic development pressures. Air and water quality standards have to be met and building permits have to be obtained, but there are many ways in which long term regional sustainability can be compromised, often through individually small but cumulatively harmful actions.

A "shopping list" of indicators that could be monitored in the context of ecosystem integrity is given in Table 1 (Munn 1990). This list is far from complete but it does indicate the complexity of the task, particularly in view of the fact, as mentioned earlier, that integrity is a moving target. Recognizing that each region may include some unique features, there are nevertheless some key questions that should be asked when trying to set monitoring priorities:

1. What are the most critical limiting factors for sustainability? Perhaps the region is already short of water, waste disposal sites, or wilderness areas.
2. In what ways is ecosystem integrity already threatened?
3. Are there likely to be any irreversible trends if current development pressures increase, and if so, how much time is available to take remedial action?
4. What are the interactions (interdependencies, conflicts of interest) amongst the various sectoral interests in the region?
5. What kinds of surprises/discontinuities might occur over the next 30 to 50 years to shock the system? For example, what would be the socioeconomic impacts in the Great Lakes Basin of a drop of 2m in lake levels, or of a doubling in population?
6. What kinds of questions will the monitoring system be expected to answer? Noss (1990), for example, has suggested the following in the context of diversity but equally applicable to ecosystem integrity:
   - Is the ratio of native to exotic range grasses increasing or decreasing?
   - Is the average patch size of managed forests increasing or decreasing?
   - Are populations of neotropical migrant birds stable?
   - Are listed endangered species recovering?

Table 1. Indicators that could be monitored in the context of ecosystem integrity and sustainability (Munn 1990).

A: General Indicators

1. State of health of natural ecosystems (primary productivity, efficiency of nutrient recycling, species diversity, population fluctuations, prevalence of pests, etc.) (Rapport 1989)
2. Resilience indicators (biodiversity, spatial patchiness, etc.)

B: Indicators of Threats to Ecosystem Integrity

1. Increasing population (especially urbanization)
2. Increasing energy consumption (type, per capita and total amounts, energy required to produce a unit of manufactured goods, etc.)
3. Increasing consumption of water (per capita and total)
4. Rates of depletion of renewable and nonrenewable resources (forests, prime agricultural land, mineral reserves, etc.)
5. Increasing amounts of wastes
6. Transportation indicators (number of cars, kilometers of road, number of airline passengers, etc.)

C: Indicators of Reduced Threats to Ecosystem Integrity

1. Increasing output of production per unit of natural resources used (both renewable and nonrenewable)
2. Increasing extent of recycling efforts
3. Conservation of scarce or highly valued resources (endangered species, wilderness, historical monuments, etc.)
4. Declining sales of pesticides and herbicides
5. Declining economic and energy subsidies given to the natural resource sectors
6. The degree of citizen involvement in "environmentally friendly" actions

There is, of course, a very rich literature on ecological indicators. See, for example, Steedman (1988), Rapport (1989), Cairns *et al.* (1991), Reynoldson (1991) and Environmental Monitoring and Assessment Vol. 15, issue no. 3 (Piekarz 1990). This store of knowledge will be of great use in designing specific systems.

Concluding Remarks

Some tentative ideas have been explored in this paper on the problem of designing a system to monitor ecosystem integrity. However, these ideas still need to be translated into practical operating systems. In this connection, a few recommendations on desirable future directions can be given:

1. Natural and social scientists should collaborate in the design and execution of a couple of long term pilot studies of ecosystem integrity in catchment areas that include human settlements. This is not such a far-fetched idea as it may seem. Note, for example, the IFGYL (International Field Year on the Great Lakes) of the 1970s, which was an enormous undertaking, involving hundreds of scientists and many planning meetings. In retrospect, IFGYL was a great success, but it lasted only a year, and it did not include measurements of indicators of human development. Experience has shown that such studies (note also those at Hubbard Brook and the Experimental Lakes Area of Northern Ontario) are beneficial in two ways. They provide data sets that can be used for testing models for many years to come. They also lead to refinements in monitoring systems, the net outcome being that more cost-effective approaches can be found, which may be applied elsewhere. (In monitoring the geophysical environment, for example, ways have been found to parameterize fine-grained spatial inhomogeneities.)

   Admittedly, the systems studied until now have been relatively simple. However, some of the methodologies developed may be transferable to areas that include human settlements. Of course, active collaboration by social scientists will be required.

2. Integrated models that combine socio-economic and ecological submodels should be developed and tested with the data collected in the pilot studies mentioned above. A first-generation model is by necessity simple but it provides a first set of indicators to be monitored. Then through successive iterations, both the model and the monitoring system gradually improve.

   Attempts to design an ecosystem model for the Great Lakes Basin are currently under way under the sponsorship of the Council of

Great Lakes Research Managers. This kind of initiative should be encouraged. (See also Munn (1989) for a proposal that an integrated acid deposition model should be developed for use by Canadian policy makers. It could also be of value in studies of ecosystem integrity and sustainability.)

3. Finally, the scientific community should exploit the opportunities given from time to time by the occurrence of step changes in some of the driving socio-economic variables, such as, introduction of new legislation, consumer products, technology, and tariffs. Given sufficient advance knowledge of the start-up time, a monitoring system could be designed to study the effects, if any, on the social and ecological fabric of the region. Of particular interest would be cases in which the region included two or more jurisdictions, each with a somewhat different response to the step changes.

ACKNOWLEDGEMENT

I acknowledge helpful discussions with Henry Regier.

REFERENCES

Allen, T.F.H., R.V. O'Neill, and T.W. Hoekstra. 1987. Interlevel relations in ecological research and management: some working principles from hierarchy theory, *J. Appl. Systems Anal.* 14, 63-79.

Allen, T.F.H. and T.W. Hoekstra. 1990. The confusion between scale-defined levels and conventional levels of organization in ecology, *J. Vegetation Res.* 1, 5-12.

Beanlands, G.E. and P.N. Duinker. 1983. *An Ecological Framework for Environmental Impact Assessment in Canada.* Inst. Resource and Env. Studies, Dalhousie University, Halifax, N.S., 132 pp.

Cairns, J., P.V. McCormick, and B.R. Niederlehner. 1991. A proposed framework for developing indicators of ecosystem health for the Great Lakes region, draft report subm. to I.J.C., 55 pp.

Edwards, C.J. and H.A. Regier. (eds.) 1990. Proc. Workshop, an Ecosystem Approach to the Integrity of the Great Lakes in Turbulent Times, Spec. Pub. 90-4, *Great Lakes Fishery Commiss.*, Ann Arbor, Mich., 299 pp.

Francis, G. and H.A. Regier. 1991. Challenges to governance from an ecosystem approach: examples from the Great Lakes, Preprint, Seminar on Ecosystem Approach to Water Management, Oslo, Norway, ECE, Geneva, Switzerland, 10 pp.

Harris, H.J., P.E. Sagar, H.A. Regier, and G.R. Francis. 1990. Ecotoxicology and ecosystem integrity; the Great Lakes examined, *ES&T*. 24:598-603.

Holling, C.S. 1973. Resilience and stability of ecological systems, *Ann. Rev. of Ecology and Systematics*. 4, 1-23.

IJC 1989. Report of the Science Advisory Committee, IJC Great Lakes Regional Office, Windsor, Ont.

IUCN 1991. *Caring for the World: a Strategy for Sustainability*. IUCN, Gland, Switzerland, 135 pp.

Kay, J.J. 1991. The concept of ecological integrity, alternate theories of ecology, and implications for decision-makers, Preprint, Canad. Envir. Advisory Council, Ottawa, Canada, 36 pp.

Mooney, H.A., E. Medina, D.W. Schindler, E.D. Schulze and B.W. Walker. (eds.) 1991. Ecosystem Experiments, *SCOPE 45*. John Wiley, Chichester, U.K.

Munn, R.E. (ed.) 1989. The case for integrated models of acidic deposition in Canadian environmental policy making, *Env. Monog.* No. 10, Inst. for Env. Studies, U. of Toronto, 57 pp.

Munn, R.E. 1990. Towards sustainable development, *Proc. Symp. on Managing Envir. Stress*. U. New South Wales, Australia, pp. 11-19.

Noss, R.F. 1990. Indicators for monitoring biodiversity: a hierarchical approach, *Conservation Biology*. 4, 355-364.

Piekarz, D. (ed.) 1990 Ecological indicators of the state of the environment, *Env. Mon. and Assess.* 15, 213-314.

Rapport, D.J. 1983. The stress-response environmental statistical system and its applicability to the Laurentian Lower Great Lakes, *Stist. J. of the U.N.* 1, 377-405.

Rapport, D.J. 1989. What constitutes ecosystem health? *Perspectives in Biol. and Medicine 1*. 121-132.

Regier, H.A. 1989. Can we get there from here? Preprint, *The Couchiching Conference*. Aug. 1989, 12pp.

Regier, H.A. 1991. Indicators of ecosystem integrity, Preprint, *Int. Symp. on Ecolog. Indicators*. Fort Lauderdale, Fla., 23pp.

Reynoldson, T.B. 1991. The development of ecosystem objectives as part of an ecosystem approach to the management of human impact on the Laurentian Great Lakes, Preprint, *Seminar on Ecosystems Approach to Water Management*. Oslo, Norway, May 1991, 8pp.

Steedman, R.J. 1988. Modification and assessment of an index of biotic integrity to quantify stream quality in southern Ontario, *Can. J. Fisheries and Aquatic Sci.* 45, 492-501.

Steedman, R.J. and H.A. Regier. 1990. Ecological basis for an understanding of ecosystem integrity in the Great Lakes Basin. In: Proc. Workshop on an Ecosystem Approach to the Integrity of the Great Lakes in Turbulent Times, Great Lakes Fishery Commission, Ann Arbor, Mich., Pub. 90-4, pp 257-270.

UNESCO 1988. Operational guidelines for the implementation of the World Heritage Convention, UNESCO. Paris, 28 pp.

National and Regional Scale Measures of Canada's Ecosystem Health

I.B. MARSHALL, H. HIRVONEN and E. WIKEN
State of the Environment Reporting
Environment Canada, Ottawa, Ontario

Introduction

This chapter outlines initiatives of State of the Environment Reporting to identify national and regional scale indicators of ecosystem health. These indicators are being developed within an ecological framework that allows the consideration of issues ranging in scale from local to continental or global.

The basis for this initiative stems from the Sustainable Development thrust of the Brundtland World Commission on Environment and Development Report (Our Common Future), the subsequent National Task Force on Environment and Economy recommendations, and the federal government response, the Green Plan. The release of *A Framework For Discussion On The Environment*, by the federal government in March 1990 began a public dialogue on a comprehensive national environment strategy and action plan: The Green Plan (Environment Canada, 1990). The final "Green Plan" report released in December 1990 represents "the government's commitment to new policy thrusts to manage our resources prudently and to encourage sensitive environmental decision making" (Government of Canada, 1990).

One cornerstone of the Green Plan is the reporting, on a continuing basis, of the State of Canada's Environment. The key element of significance to the workshop is the Green Plan's commitment to establish a long term State of Environment (SOE) ecosystem monitoring and assessment capability. SOE Reporting and the Department of Environment are currently in the initial planning stages of the strategic direction and implementation of such a program.

SOE Reporting

The State of the Environment Reporting (SOER) program has four related initiatives:

(1) *SOE Reporting and SOE Products*: will respond to the public demand for credible, timely and accurate information on the state of the Canadian environment to enable Canadians to make environmentally sensitive decisions.

(2) *Ecological Monitoring*: has as its objective to ensure that adequate spatial and temporal ecological data and analyses are available to address sustainable development issues, and support comprehensive SOE reporting and long term environmental forecasting. Essential to ecological monitoring is the selection and development of ecosystem spatial frameworks to promote understanding and improved decision making.

(3) *Environmental/Ecological Indicators*: strives to develop and publish indicators that are analogous to economic and social indicators used daily by governments, business and private citizens in their decision-making. The indicators will give a representative measurement of the state of the environment and allow an estimation of the relationship between ecosystem factors and economic development. They serve as the key to transform and summarize data after they have been collected for a sector specific purpose into understandable and meaningful information which can be communicated effectively.

(4) *National Environmental Information Network*: is to be established by 1994, as part of the Green Plan initiative. The network will provide one-window access directly to "SOE Information" and by referral to sector-specific environmental and resource information used by SOE Reporting but held by the responsible agencies — federal departments, provincial and territorial governments, universities, private sector and voluntary groups. The network is to be used to make environmental information more accessible to the public.

The focus of this chapter complements SOE Reporting's efforts to develop indicators of Canada's ecosystem health and a sound ecological monitoring framework.

Integrated Monitoring

Background

"We never knew what we had until it was gone"

This can be an epitaph for ecological mismanagement worldwide. Do we wait for our state of the environment to become critical before action takes place? Or do we employ the same common sense to the ecosystem that we increasingly apply to ourselves. We know that our health benefits greatly from monitoring the condition of our body, watching what we do and how we do it and adjusting accordingly if our health is threatened. Canadians realize that this same type of prudence in preventive monitoring must be conferred on our life supporting ecosystem. Otherwise, segments of our ecosystem will increasingly require intensive care with sometimes irreversible deleterious results.

Our expansive and diverse natural resource base has enhanced our economic and social well being. Paradoxically, it has also undermined our ability to monitor the health of our ecosystems. The diversity and, often, the remoteness of resources/ecosystems all mean that monitoring has been a costly and complex undertaking even for single resource issues.

The main catalysts for new initiatives in monitoring have been the macroscale ecosystem problems like acid rain and global climate change. They required fresh and innovative methods to furnish the information sought on a host of questions being posed by concerned individuals and groups, and to provide alternatives and solutions that would fairly integrate varied social, economic and environmental goals. Several countries have already refined their initial efforts in ecological monitoring as with the Environmental Mapping and Assessment Program (EMAP) in the United States and with the Integrated Monitoring Program administered through the ECE, in Europe.

Objective and Program Needs

The objective of the Ecological Monitoring Program is to provide authoritative spatial and temporal data and analysis to sustain and promote comprehensive SOE reporting and indicators. More specifically, the program needs are to:

- establish a long term State of the Environment monitoring and assessment capability to study resources at risk, ecosystem response and the

impact of major disruptions to ecosystems. This is a Green Plan commitment which is to be in place by 1993;
- address data gaps in the development of indicators for National SOE reporting purposes. The Green Plan commits the Government of Canada, by 1993, to release, on regular basis, a comprehensive set of indicators that measure Canada's progress in achieving our environmental goals; and
- address the requirements of SOE reporting. In 1996, the third national SOE report will be released.

By melding and improving the use of existing sectoral monitoring programs, these commitments can be partially addressed. However, by design, these programs are very singular in purpose and focus on a particular component of the ecosystem. The determination of ecosystem state is not possible by sectoral monitoring alone. Impacts on ecosystem integrity and associated responses require a comprehensive monitoring approach that considers all of the important ecosystem components in the context of the ecosystem as a whole.

The Main Program Elements

Ecological Monitoring for SOE reporting purposes must consider at least three distinct elements to address the above needs:

1. Sectoral Monitoring and Frameworks
Coordination and improvement of sectoral monitoring activities to better feed requirements for SOE reporting. The development of preliminary indicators has largely been based on the knowledge gained from the existing monitoring and survey activities.
2. Ecosystem Monitoring and Frameworks
Establish a comprehensive long term ecosystem monitoring network. Crucial indicators of an ecosystem are monitored, spatially and temporally, in such a manner as to allow for a holistic, rather than disjunct, view of cause/effects of ecosystems. The network will also address the baseline needs for trend analysis and for predictive modelling, both of which would contribute to enhanced national SOE reporting. The focus is thus to develop and implement a national monitoring network that incorporates, at a minimum, representation of the major ecosystems of Canada. This network would build on and augment information based on existing sectoral surveys. It should not be perceived as a substitute for ongoing programs designed to meet objectives other than SOE Reporting.
3. Ecosystem Forecasting and Modelling
Identification of ecosystems under stress, detection of emerging problems,

and a verification of the effectiveness of environmental programs to support long term environmental forecasting. Both the ecosystem monitoring network and appropriate sectoral monitoring programs will serve as the basis for cooperative research and predictive modelling that will aid in long-term forecasting. It will also assist in identifying and detecting the need for changes in criteria and indicators.

Guiding Principles

There are a number of principles that will guide the ultimate design, implementation, operation and presentation of results from the proposed ecological monitoring program. These include the:

- development and maintenance of departmental and external partnerships is essential;
- maintenance of national perspective (and regional ones of national significance) should always be kept in mind;
- existing sector monitoring and emphasize the value added to these existing programs;
- capitalization of existing experience gained both from international initiatives and from major issue monitoring programs such as that in place for acid rain;
- promotion of the development and implementation of national and international standards for indicators;
- public access to data and information is important;
- provision of a sound basis for focusing discussion by decision makers on past, present and potential future issues of national and regional concern; and
- tracking of progress in response to actions taken, and in areas where further action may be required.

Regional/National Scale Ecological Indicators: What are They?

Ecological indicators represent the condition of ecosystems or their components (air, water, land, biota and people). They also depict ecological exposure or response relationships to stressors, anthropogenic or otherwise. Regional ecological indicators, in the planning context, serve to guide and assess the effects of regional and national policy and program initiatives. They also serve to measure trends in the state of "health" of regional ecosystems and contribute to national overviews of the country's state of the environment, by characterizing conditions, types and locations of change.

General Categories of Indicators

In the broadest sense, ecosystems consist of varying mixtures of "man" and his environment, ranging from natural ecosystems through to man-modified ecosystems.

Three general types of indicators are identified:

- **stressor indicators:** primarily socio-economic, demographic and regulatory compliance measurements. Examples include pesticide applications, pollution emissions, land use and logging. Forest fires and floods are examples of natural stressors.
- **exposure indicators:** show whether ecosystems have been exposed to pollutants, habitat degradation or other causes of poor condition. Examples include ambient pollution concentrations, acidic deposition rates, toxic bioaccumulation, clear cut areas resulting from logging.
- **response indicators:** indicate the response of ecosystems to stressors. Examples include indices of community structure and diversity, fish tumours, habitat loss, species loss, lake pH change.

Stressor indicators tend to be focussed on the activities of man. To some degree, exposure indicators do as well. In contrast, response indicators center on environmental items. The concept of ecological stress, exposure and response indicators generally follows that described by the EPA in the Environmental Monitoring and Assessment Program (US EPA 1989).

"Everything is in compliance, but the patient is dying"

The objectives of using ecological indicators is quite different from that of enforcement and compliance monitoring normally conducted by regulatory inspectors. Compliance monitoring aims at identifying evidence of pollution attributable to individual polluters. But, as with a patient, we can monitor a series of infections, none of which exceeds dangerous levels, but the patient is still dying from the overall weakening of the body. Ecosystem monitoring focuses on sets of indicators to establish the cumulative impact and causes of all sources of pollution and disturbance harmful to ecosystem health.

Selection Criteria for Indicators

In general, desirable criteria of regional scale ecological indicators include:

1. **Clarity.** No matter how scientifically credible, if the indicator(s) is not understood by the public and by decision makers, it is of limited

value. The development and analytical process can be as technical as necessary but the final indicator(s) must be presented in an understandable manner.

2. **Scientific Credibility.** The process of integration and extrapolation involved in the development of regional scale ecological indicators must be available for scrutiny. As well, the databases used should be scientifically defensible. A quantitative approach is preferred when feasible; however, a qualitative approach may be more appropriate as the determination of ecological linkages and significance in quantitative terms is, at the best of times, fraught with uncertainty as the level of generality increases.
3. **Technical Feasibility.** The indicator should be amenable to regional trend analysis and derived from parameters suited to monitoring. Avoid the untenable wish list. Time and cost factors do not allow for the accumulation of limitless databases. Understand the technical constraints behind indicator development. This does not preclude the necessary research into indicator development. Quite simply, there is a need for certain indicators that are not yet available. Research is an essential component of any continuing initiative into the development and use of ecological indicators.
4. **Early Warning Capability.** An ecological indicator should provide a picture of not only the existing ecosystem conditions and the trend of these conditions but *indicate* possibility of degradation before adverse reactions occur.
5. **Spatial Representation.** The interpretive limits of broad scale indicators must be understood. Indicators at scales of 1:7 million or 1:10 million serve no function to provide trend information required at scales of 1:100,000. Site specific indicators, in turn, cannot be directly translated into a regional context. *For regional scale indicators the interpretive value must remain within the regional context.*
6. **Flexibility.** Flexibility applies to methodology, data resolution, and to the application and modification of the indicators themselves as new scientific information becomes available.
7. **Issue oriented.** Indicators must be relevant to identified national and regional issues of concern.

Although selection criteria have been identified, they will not preclude disagreements about the choice of indicators. Value judgements will be prevalent in the process of identifying sets of representative indicators to monitor trends in stress, exposure and response. Differing value judgements on significance may reflect the differing priority given to an environmental issue in various regions of Canada or within regions.

Reporting Framework

Trends in Demand for Information

There are significant new demands for environmental information and related management systems to reflect the relationships between sectors and socio-economic information. The traditional means of gathering information — by resource sectors such as forestry, energy, wildlife, or by medium — air, land, water can no longer cope with the complexity of today's issues. Reporting frameworks will have to take into account the following trends in demand for the general public and decision makers (Crain 1991):

1. consistent information covering many disciplines (vertical integration) over larger contiguous areas (horizontal integration);
2. a spatial component is essential to cope with the complexity and scope of today's issues. (The focus is increasingly geographical rather than statistical);
3. more information on change rather than the current state. This, in turn, focuses on the need for more rapidly updated information and consistent historical time series.

The trends have major implications for the types of support systems that will be required to deliver this type of information, particularly the demands for timely, functional information for policy decision makers.

The Spatial Context

National and regional indicators cannot be developed in isolation from consideration of issues of local, continental or global concern. They must have an ecological and spatial context. Certainly, local problems will overlap with regional and even national concerns, and cumulatively converge to create global concerns. To develop appropriate indicators, it is important not only to understand the linkages among the various spatial scales, but also the major ecological processes associated with each scale. Insight is obtained into the kind of concerns that effect each level and which ones can be aggregated from a more detailed to general level of display (Hirvonen 1990).

Figure 1 illustrates the general relationships among the various spatial scales and the major ecological influences associated with some of the more significant environmental issues of concern (Hirvonen 1990). Overlaps will occur which simply emphasize the relationships that exist among the various ecological scales.

SPATIAL SCALE	MAJOR ECOLOGICAL INFLUENCE	ECOLOGICAL CONCERNS	POSSIBLE ENVIRONMENTAL INDICATORS	TARGET CLIENTS
GLOBAL	·high level air circulation ·planetary hydrological cycles	·climate warming ·deforestation ·desertification	·CO_2 levels ·loss of world forest land	·public ·politicians ·world environmental organizations
CONTINENTAL	·tropospheric climate major vegetation zones ·major physiographic regions	·long range transport of toxics ·wildlife migration ·biodiversity	·quality of "pristine areas" ·amount of protected area	·public ·federal decision maker ·industrial opinion leaders
REGIONAL	·regional climate ·physiography vegetation communities ·watersheds	·atmospheric deposition ·wetland loss ·biodiversity ·pesticide runoff	·protected areas ·pollution emissions ·vegetation community & structure ·fishing closures ·presence/absence of wildlife species	·public ·federal & provincial decision makers ·industrial manager ·nongovernment organizations
LOCAL	·soils & landforms ·vegetation ·land use practices	·accelerated erosion ·fish and wildlife habitat loss ·water pollution ·pesticide runoff ·waste disposal	·soil loss ·site quality ·amount of waste disposal ·toxic levels in water & ·beach closures ·dead fish	·farmer ·fisherman ·scientist ·project manager ·local planner ·public

Figure 1: General relationships among ecological spacial scales, indicators and target clients.

The important subject area of linkages among indicators and between indicators and human activity, at various scales, will not be addressed in this chapter. Such linkages are the object of ongoing investigations and strategic planning for State of Environment Reporting.

Ecological Perspectives

SOE Reporting is currently looking at the ecological classification of Canada as the main spatial framework for reporting on the environmental health of the country (Wiken 1986). This classification divides Canada into 15 ecozones which can be further subdivided into 45 ecoprovinces, 177 ecoregions and over 5,400 ecodistricts. The ecological units will serve as the main reporting frameworks. Only the terrestrial units have been defined. Major aquatic systems such as the Great Lakes, Arctic Ocean or coastal and estuary systems need to be added.

The ecological units for national reporting purposes will also serve as the planning framework for a comprehensive national ecosystem monitoring network. At a minimum, monitoring sites must adequately cover the major ecozones of Canada. Actual network design would depend on the nature of a particular ecozone and on existing resource monitoring.

Challenge to Information Management and Analysis

In practice, developing a set of indicators or measures of Canada's ecosystem health has proved to be a complicated process. Recent experience from the SOE Reporting project to develop a preliminary set of indicators for the 1991 G-7 meeting has shown that we are constantly faced with systemic and interpretational problems, and changes in scale and relationships between different scales. For example:

- information exists for some parts of the country, or for some components of the environment, and not others;
- data exist for limited periods of time, thus precluding analysis of trends;
- in many instances, measurements or definitions are not standardized, so that one region cannot be compared with another; nor can data be compared from year to year;
- new techniques and advances in methodology compound problems caused by the lack of standardized measurement;
- in many cases, the data are in raw form, i.e., they have not been analyzed or otherwise processed into meaningful information that can serve as the basis for a national or a significant regional assessment. In addition, there are severe methodological difficulties in aggregating some complex data sets into a representative national overview; and

- there is inadequate auditing of the quality and relevance of the data collected (Indicator Task Force, Environment Canada 1991).

Current Activities

Partnerships and Workshops

The first guiding principle of SOE Reporting is to develop and maintain partnerships with data holders, experts and potential users within federal and provincial departments and agencies and academic community. To achieve this, a series of regional workshops on ecological monitoring and development of national and regional ecological indicators has been initiated. Discussions will take into account the guiding principles, spatial framework concerns, and the selection criteria previously outlined.

Each workshop will produce a report outlining, from the perspective of the region, the major issues, viable regional indicators of these issues, the current monitoring capability to address these issues, the identification of data gaps, and approaches/suggestions to regionalize the data. The reports, from each of the regional workshops, will form the basis for a national strategy and framework by the SOE Reporting Organization. The results will be used to address the development and implementation requirements of a national indicator and monitoring program for national and regional environmental issues.

Atlantic Region Pilot Workshop

The first workshop on ecological monitoring and indicator development was held in February, 1991, at Halifax. Under the direction of the SOE Headquarters organization, the workshop was managed, under contract, by Dalhousie University. The local planning and contract authority was the DOE Regional Office (Inland Waters). Participation was kept to a maximum of 30, representing federal and provincial resource agencies and academics based on recommendations from regional agencies.

Broad preliminary conclusions from the workshop results include the:

- endorsement of the "ecozone" as a SOE reporting framework;
- confirmation that multi-agency participation is essential at the planning and operational levels;
- preference for only one agency to provide general direction and coordination of ecological monitoring and indicator initiatives;
- confirmation that comprehensive ecosystem monitoring and assessment sites, along with sectoral monitoring (e.g., water quality net-

work) to measure the current state and trend of the health of ecozones;
- confirmation that regional issues are acid rain, toxic chemicals, land use, ozone, climate change, and forestry practices; and
- suggestion that the Atlantic Maritime ecozone would need 2-3 intensive ecosystem monitoring sites.

Future Activities

The same workshop format will be followed for the remaining regions across Canada. Dalhousie University will synthesize highlights of workshop discussions and prepare a draft report for review by all participants. A final draft will then be submitted to SOE Reporting and all participants. Participants will also receive the final national overview document.

REFERENCES

Environment Canada. 1990. *A Framework for Discussion on the Environment.* Environment Canada. Ottawa, Ontario.

Government of Canada. 1990. *A Report On The Green Plan Consultations* (August, 1990). Ottawa, Ontario.

Government of Canada. 1990. *Canada's Green Plan: Canada's Green Plan For A Healthy Environment.* (December, 1990), Ottawa, Ontario.

Hirvonen, H. 1990. Development of Regional Scale Ecological Indicators: A Canadian Approach. *State of Environmental Indicators* (SOE Report No. 91-1), State of Environment Reporting, Ottawa, Ontario.

National Task Force On Environment An Economy. 1987. Report of The National Task Force On Environment and Economy. Submitted to the Council of Resource and Environment Ministers, Ottawa, Ontario.

United States Environmental Protection Agency. 1989. *Environmental Monitoring and Assessment Program: Overview.* Office of Research and Development, U.S.E.P.A., Washington, D.C.

Wiken, E.B. (compiler). 1986. Ecozones of Canada. Ecological Land Classification Series No. 19. *Lands Directorate,* Environment Canada, Ottawa, Ontario.

World Commission on Environment and Development. 1987. *Our Common Future.* Oxford, Oxford University Press.

National Environmental Monitoring: A Case Study of the Atlantic Maritime Region

N.L. SHACKELL, B. FREEDMAN and C. STAICER
Department of Biology and School for Resource and Environmental Studies, Dalhousie University, Halifax, Nova Scotia

Abstract

A framework is being developed for a national environmental-monitoring program for Canada. Two classes of monitoring sites are suggested for each of the 15 terrestrial ecozones of Canada: (1) intensive sites in protected areas such as national parks, for baseline reference and for study of functional and structural ecology, and (2) extensive sites throughout the ecozones to determine regional effects of human activities, and to gain an overview of changes in the ecological character of the landscape. The reporting of national environmental monitoring will be coordinated by the State of the Environment Reporting Organization of Environment Canada. An obvious constraint to reporting will be the coordination and interpretation of data from different agencies. Initial steps in the proposed framework are to identify the major sources of stress in particular ecozones, and deficiencies in the available environmental information. The Atlantic Maritime ecozone is used to illustrate use of the proposed framework. Most of the extensive data can be obtained from existing monitoring programs of cooperating sectoral agencies. In this region, chemical aspects of air and water are more thoroughly monitored than are biological aspects of the ecosystem. Data describing ecosystem structure and function of intensive sites are especially deficient. The lack of biological indicators of environmental quality is important, since these are the most relevant indicators of ecological integrity.

Introduction

The ultimate goal of environmental monitoring is to avoid a deterioration of ecological integrity, or of environmental quality. In response to increasing concerns about these issues, the Government of Canada is considering a national program to monitor environmental conditions. The broad objectives of this program are to: (1) detect changes in environmental conditions on regional and national scales, (2) develop and interpret regional-and/or national-scale indicators of environmental change, and (3) predict future changes.

The present document suggests a framework for a national environmental monitoring program (Freedman *et al.* 1992). The Atlantic Maritime region is used as a case study to conceptually illustrate use of the framework. Through library research, interviews with regional scientists, and a workshop sponsored by State of the Environment Reporting Organization of Environment Canada, we identified deficiencies in current sources of environmental information, and suggest the need for additional indicators of ecological integrity, most of which are biological.

Environmental Monitoring and Ecological Integrity

The purpose of monitoring is to detect changes in ecological integrity, using appropriate indicators. The consequences of monitored changes are evaluated on the basis of a cumulative knowledge of ecological principles, gained from research. For example, changes in the rate at which natural forest is being clearcut and converted to silvicultural plantations can be monitored. The consequences, for ecological integrity, are evaluated using an understanding, however incomplete, of effects on biomass, productivity, soil and streamwater chemistry, watershed hydrology, biodiversity, and global change.

There are many indicators of environmental change. The issue is to interpret such change using ecological principles. Ideally, composite indices of ecological integrity could be made, which would ease presentation to a wider audience. However, at the present state of development of ecosystem science, any composite index of ecological integrity will be controversial, because of difficulties in assigning "value" to each variable. Nevertheless, as ecological knowledge aggrades, the application of the concept of ecological integrity will become easier. In the interim, enough is known about trends in stressed ecosystems to adequately identify ecosystems with greater integrity as being relatively: (1) resilient to stress, (2) biodiverse, (3) structurally and functionally complex, and (4) a component of a natural sere (i.e., successional sequence) that is stable over the longer term (Odum 1985, Schindler 1987, 1990, Freedman 1989, Woodley 1990).

Indicators of ecological integrity range from cellular, organismic, population and community, to landscape levels, all of which can respond to stress. Response to increasing stress can vary from simple monotonic declines, to complex curvilinear declines. The nature of decline of ecological integrity is, in part, dependent on the resistance to stress (Figure 1).

There may be uncertainty in inferring the causes and consequences of environmental change. Nevertheless, ecological monitoring and research can be used to identify significant risks to ecological integrity, and to prevent (or less desirably, to mitigate) hypothesized damages, if decision makers are convinced by the cause-and-effect evidence.

Stressor, Exposure, and Response Indicators

For the purpose of state-of-the-environment reporting, indicators of changes are classified as (*sensu* Hunsacker and Carpenter 1990):

1) STRESSOR, caused by pollutants, pesticides, climate change, forest harvesting, etc.. STRESSOR indicators are mostly associated with human activities, but can also include natural sources (e.g., forest fire, hurricanes, volcanic eruption, etc.);
2) EXPOSURE, of ecosystems to STRESSORs. Exposure is the intensity of stress experienced at a point of time, as well as the accumulated dose over time. Examples are concentrations or accumulations of toxic substances, rate of habitat loss, etc.; and
3) RESPONSE, of organisms and/or ecosystems to EXPOSURE to STRESSORs, for example mortality, change in productivity, change in species occurrence or diversity, etc.

Framework of a National Environmental-Monitoring Program

The Canadian system of Ecological Land Classification characterizes terrain on the basis of geology, soil, physical landform, and fauna and flora. The system is hierarchical, ranging in scale from the detailed Ecoelement to the broad Ecozone (Environmental Conservation Service Task Force 1982). A Canadian environmental-monitoring program would be organized by Ecozone, of which 15 have been designated for the terrestrial/freshwater parts of the nation (Figure 2). The concept of "ecozones" is merely used for reporting purposes; the actual indicator data would be obtained from cooperating provincial, federal, and nongovernmental agencies. The following are required for each ecozone (Figure 3):

1) a list of important STRESSORs affecting the ecozone, so that relevant monitoring questions can be identified,
2) a catalogue of existing monitoring databases of provincial, national, and private agencies,

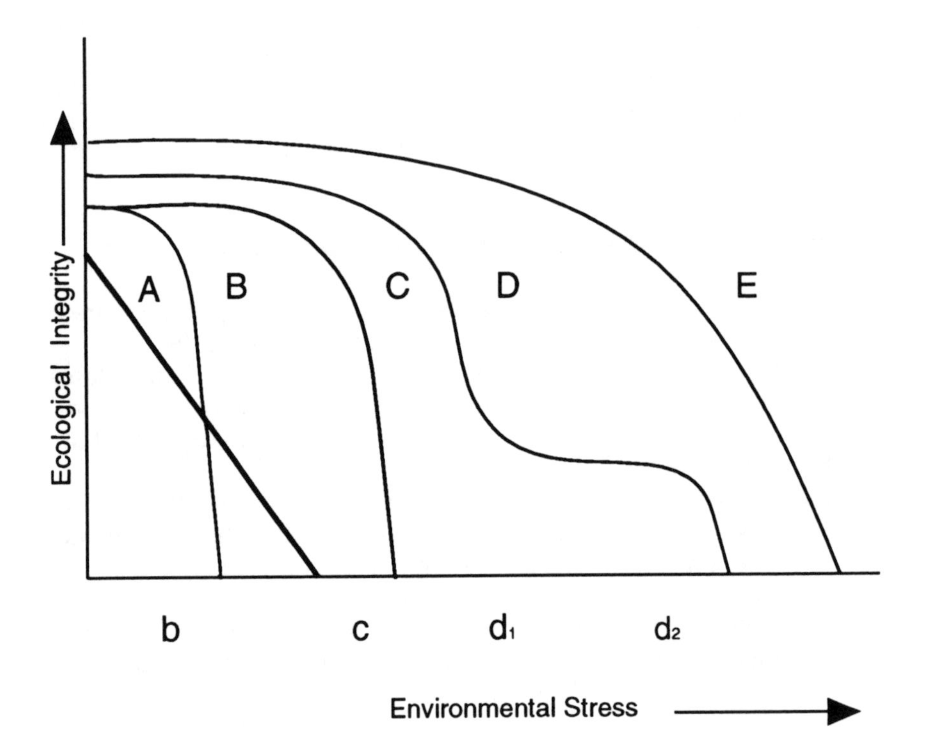

Figure 1: Ecological integrity and resistance to stress. The effects of stress on the ecological integrity of various components of the environment (e.g., organisms, communities or landscapes) depend on their resistance to stress. The curves depicted above illustrate variation in the ways that components may respond to environmental stress. Curve height represents ecological integrity, while slope represents the rate of loss of integrity in response to increasing levels of stress. Integrity is initially lower in examples A-C, and higher in examples D-E. Example A responds to increasing stress with a steady decline in integrity, but B and C begin to show a notable response only after a threshold of stress tolerance (b and c, respectively) is exceeded, with C having a higher threshold. Example D shows a complex, curvilinear response to stress, with rapid loss of integrity at one threshold (d_1), followed by relative stability (resistance), and then rapid response at a second threshold (d_2). Example E has the highest integrity initially, as well as over the longer-term. For the purpose of ecological monitoring, components A and B would be most useful as early indicators of loss of ecological integrity.

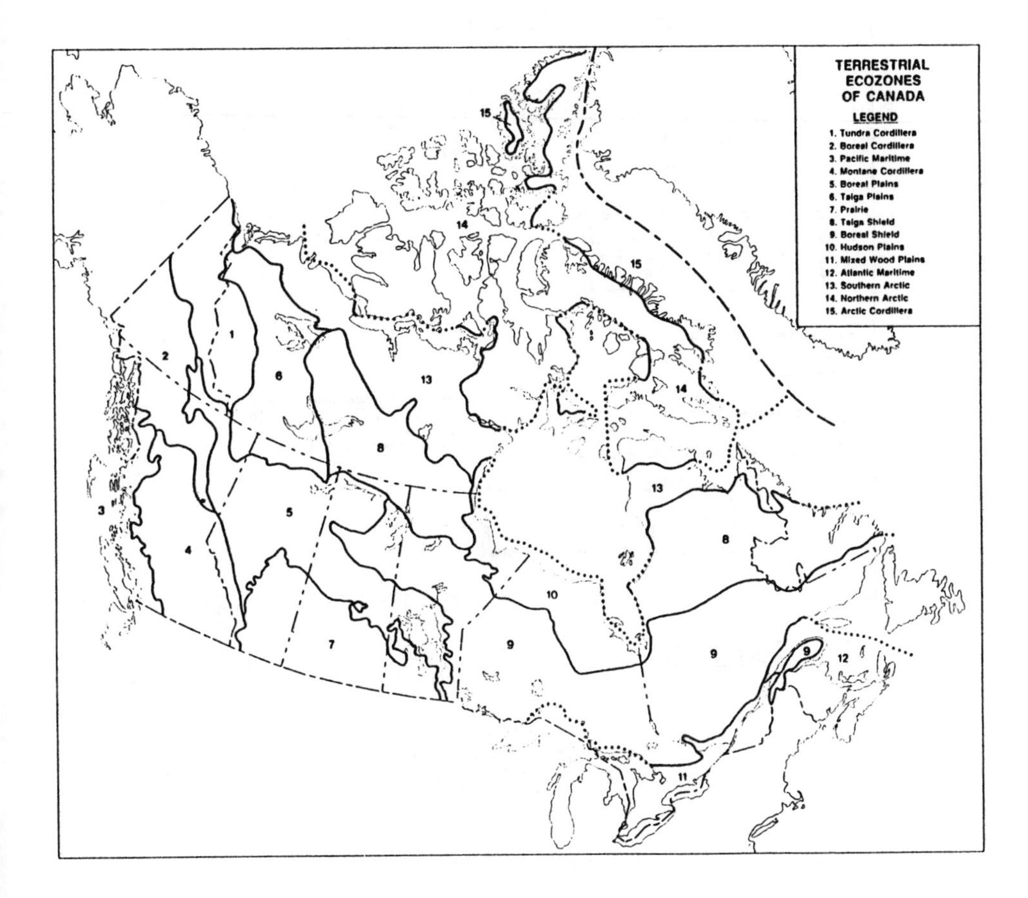

Figure 2: Terrestrial Ecozones of Canada

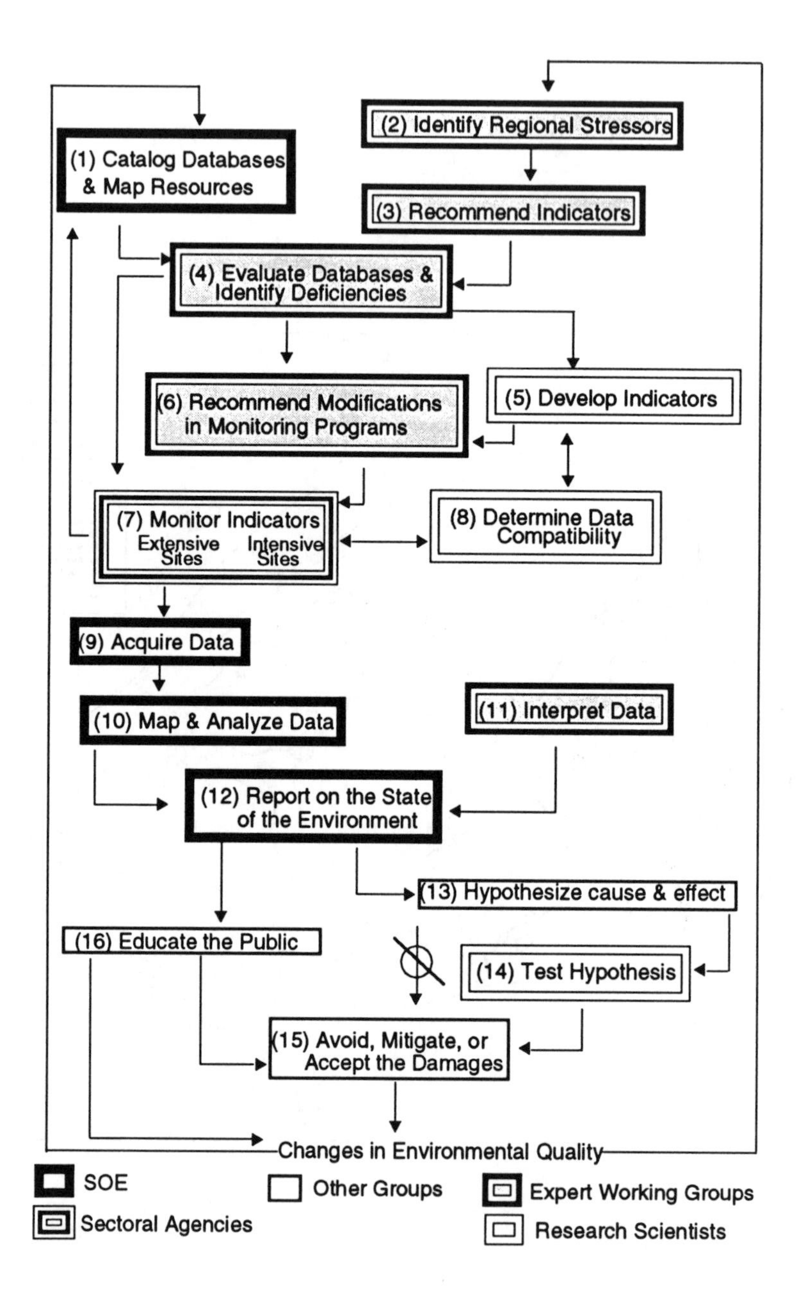

Figure 3: Functional design of the proposed national ecological monitoring program. Each box represents a different activity in the process.

3) an evaluation of the suitability of those databases for state of the environment (SOE) reporting,
4) a list of important deficiencies in the available information,
5) a list of potential indicators to address information gaps,
6) a list of potential indicators ranked on the basis of (a) how well they correlate with other indicators of environmental quality, (b) an ability to be representative on a regional scale, (c) ease of sampling, (d) suitability for sampling at all or a few ecozones or sites within the national network,
7) standardization of sampling and analytical protocol for potential indicators. Considerations include (a) temporal and spatial sampling scales that are required to allow for a statistically significant detection of change, (b) analytical procedures for chemical indicators, and (c) identification and enumeration procedures for biological indicators, and
8) methods of integration and presentation of data for the purposes of SOE reporting.

During environmental monitoring, measurements would be made within each of the 15 terrestrial ecozones of Canada. Depending on the question or issue, different spatial scales would be used. For example, global warming would affect the entire Atlantic Maritime ecozone, whereas, acidic fog is most intensive near the Bay of Fundy (Cox *et al.* 1989).

A goal of SOE reporting is to interpret regional indicators within the national context. This can impose limitations on the scale of questions asked about environmental change, particularly questions that concern relationships within specific ecosystems. Specific questions of local interest are not usually relevant to national SOE reporting. Broader scale questions that relate to basic ecological principles are more appropriate. For example, the response of voles to a change in abundance of their predators at a particular place is not a compelling question from the national SOE perspective, unless it was asked in the context of a broader change of predator populations across the ecozone or nation. An example of a broader question, relevant to SOE reporting, might be whether the biodiversity of forests has responded to changes in the silvicultural use of pesticides in the ecozone.

In the present framework for a national environmental monitoring program, two categories of monitoring sites are proposed for each ecozone. These are: (1) *intensive sites*, with detailed, multidisciplinary monitoring and experimentation, and (2) *extensive sites*, with monitoring of fewer variables. Wherever possible, intensive and extensive sites would be near existing locations within monitoring networks of air or water quality (e.g., CaPMoN, LRTAP, etc.) to facilitate determining the cause/effect relationships among indicators.

1. ***Intensive Monitoring Sites.*** Intensive, multidisciplinary studies conducted at a few sites in each ecozone are designed to:
a) detect changes in sites that are relatively undisturbed by humans, for use as a baseline reference,
b) refine currently used indicators, and develop new or better indicators,
c) detect changes in structural and/or functional attributes of monitored and/or experimentally-manipulated ecosystems, and
d) develop predictive models of ecosystem dynamics, that would allow distinguishing between natural and anthropogenic causes of change.

Intensive monitoring sites would be located in relatively remote and protected locations, such as a national park, provincial park, or some other ecological reserve. The monitored ecosystems would preferably be relatively mature and stable, so that detected changes most strongly reflect regional or global environmental changes, rather than successional dynamics. This is important because changes due to succession in young ecosystems can often overwhelm the cumulative effects due to anthropogenic STRESSORs. At each site, air, soil and water quality would be measured, as would be the diversity and abundance of the biota.

Ideally, there would be 2-3 intensive monitoring sites in each ecozone. Although redundancy is expensive, it is necessary to ensure against the catastrophic loss of particular sites.

2. ***Extensive Monitoring Sites.*** Such sites are designed to:
a) monitor the effects of extensive human activities, e.g., agriculture, forestry, mining, as compared to a reference (intensive) site or to a known historical condition, and
b) monitor changes in the ecological character of the region, as a result of changes in land-use patterns.

Much information can be obtained from national and provincial agencies, which routinely collect sectoral data on economically important indicators, in order to calculate allowable forest harvests, or hunting and fishing limits. Such data are already an important source of information for SOE reporting.

Indicators would be (and are) monitored in permanent sampling sites throughout the Ecozone, in various ecosystems (e.g., forests, agroecosystems, wetlands). Biological indicators could include the productivity, abundance, and diversity of selected species, while nonbiological measurements might include chemical quality of soil and water, hydrology, and forest and agricultural site capability.

Site Selection - Scale and Sensitivity

Extensive sampling sites can be selected by combining the spatial scales of ecodistrict and watershed, on the basis of their sensitivity to STRESSORs. The rationale for a combined spatial scale of watershed/ecodistrict is that important information could be missed if sampling is done at too large a scale (e.g., ecoregion or ecozone). Data can always be pooled at ecoregion or ecozone levels for a national SOE report.

The number and distribution of extensive sample sites is also dependent on the question being asked. For example, endangered piping plovers need only be sampled on appropriate beaches, whereas monitoring the response of vegetation to global warming requires a much more extensive sampling design.

If, for the interests of SOE monitoring, too many sample sites are proposed at the ecodistrict/watershed scale, the number can be reduced by grouping similar sites with respect to elevation, age, soil series, slope, proximity to urban areas or pollutant sources, sensitivity to STRESSORs, etc.. However, sample sites would be preferentially chosen close to air- and water quality monitoring stations (e.g., LRTAP, ARNEWS), or to be located in an ecological reserve. Aggregation of indicator measurements is advantageous, because it allows the description of relationships among indicators.

The sample design of extensive monitoring is dependent on the particular STRESSORs. Sites in the ecozone will vary in their sensitivity to particular STRESSORs, a feature that can be mapped to portray risks. The various sample sites would be located in a range of sites of low and high sensitivity. The fact that RESPONSE can often be caused by multiple STRESSORs must also be taken into account. Maps of STRESSOR sensitivities could, in turn, be overlain to combine STRESSORs (the technology of Geographic Information Systems is especially appropriate for this function). The final result could be a series of regions of low, medium, or high sensitivity to a range of STRESSORs.

Ecosystem Modelling and Trend Extrapolation

Ecosystem modelling is an important aspect of monitoring, because it allows the prediction of change at similar sites and/or in the future. It is unreasonable to propose to monitor all aspects of all ecosystems throughout an ecozone, yet ecosystem structure and function are not well understood. An important purpose of intensive sites is to elucidate ecosystem structure and function, so that relationships among STRESSOR, EXPOSURE, and RESPONSE indicators can be examined. That information could then be used to hypothesize relationships at extensive sites, where fewer variables are measured. For example, it is much easier to extensively measure a decrease in surfacewater pH caused by acid rain, than it is to monitor changes in fish communities. However, all relevant variables could be measured at intensive sites, and the relationships used to predict fish responses at comparable extensive sites where only pH is measured. Data from monitoring programs facilitate the development of models by providing initial parameter values, and by identifying key variables.

Without reference to a particular model or data set, it is difficult to address the power of extrapolation. However, general guidelines can be discussed. In order to predict a future event, we must have confidence in the existing data series. Important issues are the *accuracy* and *precision* of measurement, and the calculation of *confidence limits*.

Accuracy of measurement is the degree to which the data represent true values. Accuracy can be investigated by using different but reliable methodologies, and examining the similarity of the estimates. Precision is the replicability of methodology, or the degree to which the same methodology used in other locations or at other times will yield comparable results. Precision can be addressed by repeated measurements of the same variable, using some accurate methodology. Confidence limits are calculated for data sets having replicate measurements, in the proper statistical sense. (Recommendations for broad scale extrapolation from environmental monitoring data are outlined in Berkowitz *et al.* 1989).

A Case Study of the Atlantic Maritime Ecozone

The framework presented here is designed to be applied in each ecozone across Canada. The approach to selection of indicators that represent each region would be similar, but specific indicators might differ among ecozones. In this section, the Atlantic Maritime ecozone is used to conceptually illustrate application of the approach outlined in the framework.

A workshop was convened in Halifax in February, 1991, to discuss environmental monitoring in the Atlantic Maritime ecozone. Most of the participants were government scientists involved in environmental research and/or monitoring in the Atlantic Maritime region. The objectives of the workshop were to: (1) discuss regionalization of environmental data pertinent to SOE reporting, and (2) identify important deficiencies in the available information, and suggest monitoring requirements and other initiatives to address those deficiencies. Workshop participants were divided into three groups. Two groups discussed regional indicators measured at extensive monitoring sites. The third group discussed potential and existing environmental monitoring at Kejimkujik National Park, as an example of an intensive monitoring site.

Groups approached the selection of indicators by identifying the STRESSORs pertinent to the Atlantic Maritime region, and then evaluating whether appropriate EXPOSURE and RESPONSE indicators were available. We used the results of the workshop, combined with our own opinions, to formulate matrices of STRESSOR (vertical) by EXPOSURE and RESPONSE indicators (horizontal) (Appendix 1, see also Freedman and Shackell 1991).

STRESSORs were broadly classified, to simplify presentation and allow for future monitoring of unanticipated STRESSORs. EXPOSURE and RESPONSE indicators were selected on the basis of their ability to represent the region, following from discussions in the workshop and elsewhere. A catalogue of environmental databases in Atlantic Canada has been compiled by Environment Canada (Spencer and Neumeyer 1989). Availability of many of the indicators that we selected was based on information in that catalogue.

An intent of the workshop was to illustrate the use of our national environmental-monitoring framework, by considering the Atlantic Maritime region from an ecological perspective. The resultant matrices of indicators are relatively comprehensive, and are meant to provide SOE reporters with an overall picture of data availability and requirements. However, in an operational environmental-monitoring program, with its financial and other constraints, shorter lists of indicators would be formulated by specialized working groups. These specialists would also be responsible for the ultimate sampling designs and protocol.

Summary of the Case Study

An initial step in the design of a national environmental-monitoring program is identification of existing data appropriate for SOE reporting (i.e., are the data of high quality, and can they be used to represent the region). A consensus among workshop participants was that there is a large amount of available information relevant to SOE reporting, but there are deficiencies in the types of information being collected. Most important is a lack of data relevant to functional and structural aspects of ecosystems. Specifically, there are few data for RESPONSE indicators, and in particular, non-economic wildlife species (wildlife, in the present context, refers to any plant or animal life other than humans). Finally, it was noted that there are important impediments to SOE reporting in the coordination and assessment of data from different agencies.

There has been substantial research on the development of STRESSOR and EXPOSURE indicators of environmental quality, most of which involve monitoring nonbiological indicators, e.g., air or water chemistry, etc. There has been much less work on RESPONSE indicators, most of which are biological (see Stokes and Piekarz 1987, Sheehy, 1989). This imbalance occurs primarily because it is difficult to determine whether changes in RESPONSE indicators are due to natural variation (e.g., succession), or to changes in EXPOSURE to STRESSORs. For this reason, research is needed on the interpretation of RESPONSE indicators. Of course, research would have to be done within the context of major STRESSORs in the ecozone (e.g., acidification, global climate change, ground-level ozone, deforestation).

Given that ecological principles must be used in the operational definition of ecological integrity, or of environmental quality, the most important deficiencies of SOE monitoring in the Atlantic Maritime ecozone are related to incomplete baseline inventories of flora and fauna, and to an incomplete understanding of the functioning of ecosystems. A generic list of missing, under represented, and/or potential RESPONSE indicators in the Atlantic Maritime ecozone is presented below.

1) ***Habitat Diversity***. Landscape changes affect wildlife diversity, largely by affecting the diversity of habitats. Ultimately, management for diversity of habitats, of differing successional ages, may be easier than trying to attain population goals for all species. For example, intensive forest management affects habitats greatly, and this could be monitored across forest types through succession. The stand types could include a range of managed forests, with various types of harvest (e.g., clear-cut, selection cut) and silvicultural practices (e.g., scarification, herbicide use, etc.), and with natural forest as a reference. Beyond forestry, the loss of specific habitats over time could be measured using data from the Canadian Land Inventory (Environmental Information Systems, Spencer and Neumeyer 1989).
2) ***Ecological Reserves***. Ecological reserves are important for the conservation of (a) rare and endangered species and their habitat, (b) endangered or unusual ecosystems, and (c) representative types of ecosystems for use as reference sites. Changes in the number, area, and conservational "adequacy" of designated ecological reserves within the ecozone would be monitored over time.
3) ***Rare and Endangered Species and Their Habitat***. At present, the Canadian Wildlife Service, in conjunction with the Canadian Nature Federation and the World Wildlife Fund, periodically updates the status of endangered species to ensure the protection of such species and their habitat. These agencies have initiated a Recovery program of Nationally Endangered Wildlife (RENEW), but to 1991 only about 20% of designated endangered species (5 mammals, 7 birds, 1 fish, and 1 plant) had recovery programs. Endangered reptiles, amphibians, plants, and invertebrates are under represented, in part, because

they have less "charisma" than mammals and birds. Recovery programs and development of indicators that are appropriate in the broader ecological context, can be facilitated by monitoring a comprehensive set of endangered species.

4) ***Opportunistic Species***. Monitored changes in the abundance and distribution of "pests" (from the human perspective) indicate changes in environmental quality. As humans alter the environment, favorable conditions may be created for particular, opportunistic species.
5) ***Species Sensitive to Change***. Changes in the abundance, vigor, and distribution of widespread species that are known to be sensitive to particular STRESSORs and EXPOSUREs can be monitored as a bio-indication of environmental quality. Examples include lichens, which can be effective bio-monitors of sulfur dioxide, ozone, heavy metals, and other toxic chemicals, and plant species that are close to the limits of their distribution, and which might be expected to respond relatively strongly to climatic change.

Conclusion

Environmental monitoring is ultimately carried out to avoid a deterioration of ecological integrity. Important aspects of design in a national environmental-monitoring program relate to coordination of existing sectoral monitoring, the use of intensive (reference) and extensive (regional) monitoring sites within ecozones, and interpretation of the causes and consequences of monitored changes. At the present time, the most important deficiency in environmental monitoring in Canada is the lack of indicators of biological response to anthropogenic stressors.

ACKNOWLEDGEMENTS

This work was supported by a contract from State of the Environment Reporting Organization of Environment Canada. We thank H. Hirvonen for commenting on preliminary drafts of this manuscript. We are also grateful to the participants in our workshop discussions.

REFERENCES

Berkowitz, A.R., J. Kolasa, R.H. Peters, and S.T.A. Pickett. 1989. How far in space and time can the results from a single long-term study be extrapolated? In: G.E. Likens (ed.). *Long-term Studies in Ecology: Approaches and Alternatives*. Springer-Verlag, Inc. New York.

Burnett, J.A., C.T. Dauphine, S.H. McCrindle, and T. Mosquin. 1989. *On the Brink. Endangered Species in Canada*. Western Producer Prairie Books. Saskatoon, Sask.

Cox, R.M., J. Spavold-Tims, and R.N. Hughes. 1989. Acid fog and ozone: their possible role in birch deterioration around the Bay of Fundy, Canada. *Water, Air, and Soil Pollution*. 48: 263-276.

Environment Canada. 1988. Effects of acid rain on Atlantic Canada's inland waters. EN1-8/1-1988E. Atlantic Region State of the Environment Reporting, Conservation and Protection, Environment Canada. Dartmouth, N.S.

EPS. 1989a. National air pollution surveillance for 1987. EPS 7/AP/20, Environmental Protection, Environment Canada. Ottawa.

EPS. 1989b. National air pollution surveillance for 1988. EPS 7/AP/21, Environmental Protection, Environment Canada. Ottawa.

Freedman, B. 1989. *Environmental Ecology*. Academic Press, San Diego, CA.

Freedman, B. and N.L. Shackell. 1991. National Environmental Monitoring Program: a case study for the Atlantic Maritime Ecozone. Prepared for State of the Environment Reporting Organization, Environment Canada. Department of Biology and School for Resource and Environmental Studies, Dalhousie University. Halifax, N.S.

Freedman, B., C. Staicer, and N. Shackell. 1992. Recommendations for a national ecological-monitoring program. Prepared for State of the Environment Reporting Organization, Environment Canada. Department of Biology and School for Resource and Environmental Studies, Dalhousie University, Halifax, N.S.

Hartman, G.F. and J.C. Scrivener. 1990. Impacts of forestry practices on a coastal stream ecosystem. Carnation Creek, British Columbia. *Can. Bull. Fish. Aquat. Sci.* 223. Department of Fisheries and Oceans. Ottawa.

Howell, G. 1986. Water quality characterization fog 78 acid precipitation monitoring lakes in Atlantic Canada. IWD-AR-WQB-86-109. Environment Canada, Inland Waters Directorate, Atlantic Region. Moncton, N.B.

Hummel, M. (gen. ed.). 1989. *Endangered Spaces. The Future for Canada's Wilderness*. Key Porter Books. Toronto.

Hunsacker, C.T. and D.E. Carpenter (eds.). 1990. *Ecological Indicators for the Environmental Monitoring and Assessment Program*. EPA 600/3-90/060. U.S. Environmental Protection Agency, Office of Research and Development, Research Triangle Park, N.C.

Hussell, D.J.T., M.H. Mather, and P.M. Sinclair. 1990. Trends in numbers of tropical and temperate landbirds in migration at Long Point, Ontario, 1961-1988. *Symposium on Ecology and Conservation of Neotropical Migrant Landbirds*. Manomet, Mass.

Kuhnke, D.H. 1989. Silviculture statistics for Canada: an 11-year summary. Inf. Rep. NOR-X-301. Forestry Canada, Northern Forest Research Centre. Edmonton, Alta.

Lands Directorate. 1986. Wetlands in Canada: a valuable resource. Fact Sheet 86-4, Cat. No. En 73-6/86-4E. Lands Directorate, Environment Canada. Ottawa.

Magasi, L.P. 1988. Acid rain national early warning system: manual on plot establishment and monitoring. Report DPC-X-25. Canadian Forest Service. Ottawa.

Magasi, L.P. 1990. Forest pest conditions in the Maritimes in 1989. Information Report M-X-177. Forestry Canada, Maritimes Region. Fredericton, N.B.

Manuel, P. 1987. The failure of wetland conservation programs to accommodate peatland environments: A case study of the maritime provinces. C.D.A. Rubec and R.B. Overend (eds.). *Proceedings from Wetlands/Peatlands Symposium*. Edmonton Convention Centre, Alberta Canada.

Odum, E.P. 1985. Trends expected in stressed ecosystems. *Bioscience.* 35: 419-422.

RMCC. 1990. The 1990 Canadian long-range transport of air pollutants and acid deposition assessment report. Part 1. Executive summary. Federal/Provincial Research and Monitoring Coordinating Committee. Environment Canada. Ottawa.

Schindler, D.W. 1987. Detecting ecosystem responses to anthropogenic stress. *Can. J. Fish. Aquat. Sci.* 44: 6-25.

Schindler, D.W. 1990. Experimental perturbations of whole lakes as tests of hypotheses concerning ecosystem structure and function. *Oikos.* 57: 25-41.

Sheehy, G. 1989. Environmental Indicator Research. A literature review for State of the Environment Reporting. Technical Report Series. Report No. 7. *Strategies and Scientific Methods.* SOE Reporting Branch. Canadian Wildlife Service, Environment Canada.

Spencer, C.J. and R. Neumeyer. 1989. Catalogue of Environmental Data in Atlantic Canada. Conservation and Protection, Environment Canada, Atlantic Region.

Stokes, P. and D. Piekarz (eds.). 1987. Ecological indicators of the State of the Environment. Report of a workshop held May 27-29, 1987 at the Institute for Environmental Studies, University of Toronto, for Environmental Interpretation Division, Environment Canada.

Watt, W.D., D. Scott, and J.W. White. 1983. Evidence of acidification of some Nova Scotian rivers and its impact on Atlantic salmon, *Salmo salar. Can. J. Fish. Aquat. Sci.* 40: 462-473.

Wellings, F.R. and B. Barron. 1988. Multistage integrated forest inventory. Field specifications. Nova Scotia Department of Lands and Forests, Forest Resources and Planning and Mensuration Division. Truro, N.S.

Woodley, S. 1990. *A database for ecological monitoring in Canadian national parks.* Heritage Resources Centre, University of Waterloo. Prepared for Canadian Parks Service of Environment Canada.

APPENDIX 1

Indicators relevant to SOE monitoring in the Atlantic Maritime Ecozone. The following brief descriptions refer to indicators listed in Appendix Table 1.

Stressor Indicators

1) ***Land use*** is a generic STRESSOR. Land use can range from industrial use to natural habitat, with the former causing a greater detriment to the character of the land. Examples of land use include sites of power generating stations, transmission lines, oil refineries, steel plants, pulp mills, dams, roads, etc., and more extensive land uses such as residential development, agriculture, forestry, and mining.
2) ***Resource extraction*** refers to activities associated with agriculture, forestry, mineral extraction, mining, hunting, fishing, etc.
3) ***Industrial emissions*** refer to emissions from point-sources (e.g., power plants, pulp and paper mills, smelters, etc.) that have localized effects on the surroundings of the facility.
4) ***Climate change*** may be influenced by changes in atmospheric concentrations of carbon dioxide, methane, nitrous oxides, and other greenhouse gases. Effects could include an increase in global temperature, which could alter patterns of precipitation, species distribution, and sea level, among other effects (Freedman 1989). Climate change could also occur through thinning of the stratospheric ozone layer, which would result in an increase in harmful ultra-violet radiation reaching the surface of the Earth (Freedman 1989).
5) ***Long Range Transport of Air Pollutants (LRTAP)*** refers to atmospheric pollutants originating from sources outside the ecozone, and transported to the Atlantic Maritime region. Examples of LRTAP include acidic deposition, sulfur dioxide, ozone, etc.. Much of the ecozone has soil and bedrock types with little acid-neutralizing ability, rendering large areas sensitive to acidification (Environment Canada 1988, RMCC 1990). Ground-level oxidants have recently received attention in this area, and may play a role in the dieback of white birch along the Fundy coast (Cox *et al.* 1989).
6) ***Natural disturbance*** includes fire, windstorm, outbreaks of insect pests.

Exposure Indicators

Exposure Indicators refer to the rate of change and/or intensity of STRESSORs.

1) ***Land Use.*** The rate of change in land use gives a regional overview of the character of the landscape at different times. Relevant data are available for the Atlantic Maritime ecozone from Environment Canada (Environmental Information Systems database, Spencer and Neumeyer 1989). There are deficiencies about changes in wildlife habitat, largely because there is no systematic methodology within the region to assess habitat or to identify species at risk, especially nongame species (Burnett *et al.* 1989). However, initiatives are beginning at the provincial level (Wellings and Barron 1988). For game species and fur bearing mammals, provincial and nongovernmental wildlife agencies periodically assess habitat (Spencer and Neumeyer 1989).

   Changes in the number of roads, housing units, shopping malls, etc., are an indication of the rate of urbanization. The rate of conversion from one ecosystem type to another (e.g., agriculture to forest, wetland to agriculture, natural forest to tree plantation, etc.) offers an overview of changes in the character of the landscape. In the Atlantic Maritime region, 30% of farms were abandoned by 1960, most of which reverted to forests or wetlands (Lands Directorate 1986).

   In some areas, natural forests are being converted to intensively managed conifer plantations (related data are in Kuhnke 1989). Environmental issues include sustainability, effects on biodiversity, soil quality, surface waters, and conflicts with other uses of the land (Freedman 1989).
2) ***Industrial Emissions.*** There is no systematic monitoring of point-sources of industrial emissions in the Atlantic Maritime region. However, emissions from certain industrial sites are monitored by industry itself, or by provincial or federal agencies (e.g., ATLMON, NAQUADAT, or HYDAT databases, Spencer and Neumeyer 1989).
3) ***LRTAP Indicators.***
   a) *Precipitation.* The chemistry and volume of precipitation is being monitored throughout the region by various agencies (Lake Chemistry Data Base ref. 90, Rain Chemistry Data Base ref. 91, LRTAP ref. 92 in Spencer and Neumeyer 1989).
   b) *Air Quality.* Concentrations of gaseous atmospheric pollutants are monitored throughout the region by various agencies (EPS 1989a,b, CaPMoN ref. 101 in Spencer and Neumeyer 1989).
   c) *Oxidants.* Concentrations of ground level ozone are monitored at selected sites in the Atlantic Maritime region by Environment

Canada (EPS 1989a,b, CaPMoN ref. 101 in Spencer and Neumeyer 1989).

4) ***Chemical Use: Domestic, Agriculture and Forestry***. Data on the types, quantities, and value of the use of fertilizers and pesticides would indicate the risks of these activities. Relevant data for forestry are available from the FORSTATS database (e.g., Kuhnke 1989), but comprehensive data for domestic and agricultural use are not available or readily accessible.
5) ***Forest Management***. Environmental effects of forest management are related to the area that is annually treated in various ways. Relevant information includes the area of: forest harvested by various techniques (e.g., clearcut, stripcut, shelterwood, etc.), scarified, replanted with particular species, naturally regenerated, etc.. Data are accessible in the FORSTATS database (Kuhnke 1989).
6) ***Ecological Reserves***. Ecological reserves by area of habitat type are an index of ecological diversity in reserve. There are deficiencies in the present information, largely because the provinces have not established a methodology to assess biodiversity. However, some relevant data are available from databases of Parks Canada, provincial governments, and the Island Nature Trust in P.E.I. (recently summarized in Hummell 1989).
7) ***Habitat Fragmentation***. Some wildlife species require large and contiguous areas of forest. The degree of fragmentation has important implications for biodiversity, an index of ecosystem integrity (Schindler 1990). Data are not presently available, but they could be interpreted from airphotos.
8) ***Habitat Diversity***. Wildlife diversity is related to habitat diversity, both of which have been neglected in monitoring programs. Some relevant information can be obtained from forestry data on "cover type by maturity class." This indicator is an index of the proportion of forested land in different age classes, often by tree species, which reflects habitat availability and diversity. Such information is available from forest inventories (Kuhnke 1989).

Response Indicators

1) ***Land capability***. Land capability is an estimate of the quality of land or freshwater for forestry, agriculture, ungulates, waterfowl, or sportfish (Environmental Information Systems, ref. 2 in Spencer and Neumeyer 1989). Capability can be described in terms of soil quality, drainage, topography, food species, etc.. Information is usually obtained by interpretation of air photos, satellite imagery, and hydrological data.

2) ***Inland Waters.***
   a) *Fish community.* Changes in the distribution and abundance of fish can indicate changes in water quality (although the relationship is obscured in areas of intense fishing activity). However, presence or absence of species can indicate the state of the environment. For example, salmonids do not occur in highly acidic waters in the Atlantic Maritime region (Watt *et al.* 1983, Environment Canada 1988).
   b) *Fish health.* The occurrence of fish diseases is routinely monitored throughout the Atlantic Maritime region (Fish Health Data Base, ref.17 in Spencer and Neumeyer 1989).
   c) *Aquatic birds and other biota.* Canadian Wildlife Service and provincial agencies regularly census waterfowl throughout the Atlantic Maritime region (National Harvest Surveys ref.66 in Spencer and Neumeyer 1989). Some information is also available for other aquatic biota (Nutrient and Biological Productivity in Atlantic Region Waters ref.46 in Spencer and Neumeyer 1989).
   d) *Water chemistry.* Changes in water chemistry (e.g., pH, major ions, nutrients, dissolved oxygen, metals, etc.) can indicate: i) habitat quality, ii) effects of atmospheric deposition, and iii) effects of land use in the watershed (Freedman 1989, Hartman and Scrivener 1990). Water chemistry is thoroughly monitored in the Atlantic Maritime by federal and provincial agencies (Howell 1986, references in Spencer and Neumeyer 1989).
3) ***Wetlands.***
   a) *Water chemistry and biota.* Baseline inventories of water chemistry, vegetation, and animal species have been made by Canadian Wildlife Service to identify wetlands important to waterfowl (Maritime Wetlands Inventory ref.85 in Spencer and Neumeyer 1989). However, wetlands unsuitable for waterfowl (especially peatlands) are not as well surveyed (Manuel 1987).
4) ***Forests.*** Forest inventories and silvicultural data are available from provincial agencies and Forestry Canada (FORSTATS database, recently summarized in Kuhnke 1989). Especially relevant indicators include:
   a) *Productivity.* Forest productivity indicates the net increment in merchantable biomass.
   b) *Forest community by site.* This is an indicator of habitat diversity. Tree data are broadly available, but most other measures of habitat are deficient.
   c) *Age-class distribution.* This can be an index of the amount and diversity of clearcuts, and of young, mid-successional, mature, and old-growth habitat.

d) *Pest and disease incidence and forest decline.* The abundance and distribution of injurious forest insects and tree diseases are routinely monitored by Forestry Canada (Magasi 1990). Possible effects on trees of LRTAP are being monitored through the Acid Rain National Early Warning System (ARNEWS) (Magasi 1988). For example, near the Bay of Fundy, leaf browning of birch has been correlated with fogwater acidity and nitrate, and possibly with ozone (Cox *et al.* 1989).

e) *Soil quality.* Data relevant to changes in forest soil quality (organic matter, acidity, nutrient availability, soil biota, etc.) are not available on a regional scale. Some monitoring data are available from the Acid Rain National Early Warning System network (ARNEWS, Magasi 1988), and provincial agencies categorize some soil characteristics on forest inventory plots (E. Bailey Nova Scotia Department of Lands and Forests, pers. comm.). However, there is no longer term, regional monitoring of the effects of LRTAP and intensive harvesting and management on soil quality.

f) *Non-economic indicators.* Prominent in this category are the species composition and abundance of nongame animals and noncommercial vegetation, as well as habitat indicators (e.g., snag size and density, logs-on-the-ground, vertical profiles of foliage, habitat mosaic, etc.). Few data are available on a regional scale.

5) ***Agriculture.*** Agricultural systems are monitored in a relatively extensive program. Data are available from Agriculture Canada or Statistics Canada. Especially relevant indicators include:

a) *Production.* Changes in the productivity of crop species or in area planted may indicate a RESPONSE to STRESSORs (changes in crop area may also occur in response to economics). For example, certain species are relatively susceptible to pollutants (e.g., some varieties of tobacco are hypersensitive to ozone).

b) *Soil Quality.* The quality of agricultural lands across Canada will be monitored by the National Soil Quality Evaluation Program (K. Webb Agriculture Canada, pers. comm.). The program is in a pilot phase, but when fully operational it will report on a regional scale.

c) *Crop damage.* The loss of crops to inclement weather, insects, and disease is routinely monitored by provincial and federal agencies.

6) ***Wildlife.***

a) *Rare and endangered species.* The distribution and abundance of selected species are monitored by provincial wildlife agencies and the Canadian Wildlife Service, in conjunction with non-governmental organizations such as the World Wildlife Fund.

However, most rare and endangered species are not being investigated (Burnett *et al.* 1989).

b) *Game species.* The abundance of game species is routinely monitored by provincial wildlife agencies, and the data are readily available.

c) *Nongame species.* As noted earlier, there are very few data on nongame species. A comprehensive monitoring program requires baseline inventories of major taxa. For nongame birds, which are relatively well studied, there are three relevant sources of information: (i) the Breeding Bird Survey, (ii) the Christmas Bird Count, and (iii) the Breeding Bird Atlas. However, in some respects the quality and usefulness of those data for SOE monitoring are questionable. More useful regional information might be obtained from bird observatories located on major migration routes, for example the Long Point Bird Observatory in southern Ontario (Hussell *et al.* 1990).

d) *Habitat diversity.* Landscape changes affect wildlife diversity, largely by affecting the diversity of habitats. Ultimately, management for diversity of habitats may be easier than trying to attain population goals for all species. For example, since 1988, the Canadian Wildlife Service has been monitoring bird populations in habitats subjected to intensive forest management in New Brunswick (Forest Bird Habitat Requirements ref. 19 in Spencer and Neumeyer 1989). Beyond forestry, the loss of specific habitats over time could be measured using data from the Canadian Land Inventory (Environmental Information Systems ref. 2 in Spencer and Neumeyer 1989).

Appendix Table 1. Matrix for extensive monitoring sites in the Atlantic Maritime ecozone of STRESSORs (vertical) by EXPOSURE and RESPONSE indicators (horizontal). Each cell indicates whether the data for EXPOSURE or RESPONSE indicators are: available (A), of limited availability (a), deficient (D), or irrelevant to the STRESSOR (-). See text for descriptions of indicators (modified from more detailed matrices in Freedman and Shackell 1991).

	STRESSORS					
	Land Use	Resource Extract.	Industrial Emissions	Climate Change	LRTAP	Natural Disturb.
EXPOSURE INDICATORS						
Industrial Emissions	—	—	A	—	—	—
LRTAP Indicators	—	—	a	—	A	—
Pollutant Inputs	—	A	A	A	A	—
Chemical Use	A	A	—	—	—	—
Forest Management	A	A	—	—	—	—
RESPONSE INDICATORS						
Land Use	A	A	A	—	—	A
Land Capability	A	A	—	—	—	a
Ecological Reserves	a	A	—	—	—	—
Habitat Fragmentation	a	a	—	—	—	a
Habitat diversity	D	D	—	D	D	—
A) Inland Waters						
Fish community	—	A	A	—	A	—
Fish health	—	A	A	—	A	—
Aquatic birds, biota	—	—	—	—	a	—
Water pH	A	A	a	—	A	—
B) Wetlands						
Biota	—	a (baseline inventories) —			—	—
C) Forests						
Productivity	—	A	—	D	D	—
Forest community	A	A	—	D	A	—
Age-class distribution	A	A	—	—	—	—
Pest incidence, forest decline	—	A	—	—	—	—
Soil quality	—	a	—	—	A	—
Non-economic indicators	D	D	—	D	D	—
D) Agriculture						
Production	—	A	—	D	D	—
Soil quality	—	a	—	—	a	—
E) Wildlife						
Endangered species	a	a	—	D	D	—
Diversity - game spp.	a	a	—	D	D	—
Diversity - non-game	D	D	—	D	D	—

Monitoring and Measuring Ecosystem Integrity in Canadian National Parks

STEPHEN WOODLEY
Canadian Parks Service, Ottawa, Ontario

Introduction

Canada has a system of 34 national parks covering some 18 million hectares. The goal of these national parks is to preserve representative examples of Canada's major natural environments in a natural state. The Canadian National Parks' Act, amended in 1988, requires that maintenance of ecological integrity should be a first priority when considering a park management plan. Further, the Act requires that management plans and reports on the state of national parks be submitted to Parliament. Such legal or policy requirements to assess the state of ecosystems are increasingly common for resource management agencies. The state of ecosystems is commonly defined in terms of ecosystem integrity, ecosystem health or even biodiversity. Such terms are used in a normative sense and cannot be defined easily or fully by ecosystem science. Thus, resource management agencies have been struggling with trying to operationalize these terms. This chapter reports on the general framework for ecosystem integrity monitoring in Canadian national parks with emphasis on the choice of a suite of suitable measures from ecosystem science.

The Goals of Canadian National Park Monitoring

Monitoring measures are generally designed to be able to provide information to (1) understand, (2) detect changes and (3) *perhaps* manage or control a system. The objectives of ecosystem monitoring are usually to understand long term changes in the ecosystem, identify baseline conditions, follow a response to a specific threat, or ensure specific conditions (e.g., clean drinking water) are maintained. For the purposes of this project, the goals for monitoring ecosystems within the Canadian Parks Service are more specifically defined as follows:

1. To measure and detect changes in the ecological integrity or state of health of the ecosystem or ecosystems that a given national park has been established to protect. This understanding is required so that the park ecosystems may be protected, and timely management action taken where required. Such understanding is also beneficial for interpreting the park to visitors and for researchers conducting programs within the park.
2. To measure the effects of specific, perceived threats to the ecosystem or ecosystems that a national park has been established to protect.
3. To provide data on the state of Canadian national parks to other national and international agencies and researchers conducting state of environment reporting or monitoring large scale environmental change such as acidic precipitation and climatic change.

   It follows from the above goals, that there is not currently a satisfactory degree of knowledge about the state of health of the ecosystems protected by Canadian national parks. A first report on the state of national parks was prepared in 1990 (Canadian Parks Service) and it was considered to be a developmental exercise (André Savoie, pers. comm.). Therefore, a fourth goal of the development of a parks monitoring system is:
4. To provide a database that can be used for the preparation of a report on the state of national parks in Canada. Such reports are required by law under the amended National Parks Act of 1988.

Defining Ecological Integrity

As ecosystems worldwide are put increasingly under stress, the need to preserve the ecological processes, on which all life depends, has been widely recognized. Recently, The United Nations' World Commission on Environment and Economy recognized that human society was dependant on uses of the biosphere that were sustainable (Bruntland 1987). Human values are an integral part of the decisions to protect or rehabilitate, but the goals and objectives for such actions are often implicit or unclear. Ecosystems are often valued because of certain features they contain or functions they serve. National parks and other protected areas are set up as an expression of a set of human values, traditionally expressed as the preservation of natural ecosystems. More recently, the conservation of natural areas is being justified in terms of maintaining ecosystem integrity.

There are several published definitions of ecosystem integrity in the literature. The first can be traced to Aldo Leopold (1939), whose poetic definition sets a tone for the ethical basis for integrity, but sheds less light on more objective measures:

A thing is right when it tends to preserve the integrity, stability and beauty of the biotic community. It is wrong when it tends otherwise.

A more measurable definition is provided by Cairns (1977), who mixed both ethical and ecological elements:

Biological integrity is the maintenance of the community structure and function characteristic of a particular locale or deemed satisfactory to society.

Karr and Dudley (1981) defined integrity in more quantitative terms and even developed a quantitative index of biological integrity for streams. This index measures species composition and trophic structures against a local reference stream deemed to be healthy:

Biological integrity is the capability of supporting and maintaining a balanced, integrated, adaptive community of organisms having a species composition and functional organization comparable to that of the natural habitat of the region.

Other definitions have attempted to be completely objective in defining integrity. For example Kay (1989) defined integrity in terms of the thermodynamic properties of ecosystems.

If a system is able to maintain its organization in the face of changing environmental conditions, then it is said to have integrity.

For national parks, we propose a definition of ecosystem integrity that recognizes both ethical judgment and quantitative elements provided by ecosystem science. Ecosystems are viewed as dynamic systems that are adapted to internal stresses such as fire. However, ecosystems are not adapted to many human-caused stresses. These stresses lead to degradation and loss of integrity. Ecosystems are also communities of co-evolved species and these "native" species are not interchangeable with "non-native" species. The definition also recognizes that species exist in populations that must be kept above a minimum level if they are to persist. Finally, the definition of integrity in national parks reflects the basic purposes of national parks:

Ecological integrity is defined as a state of ecosystem development that is optimized for its geographic location, including energy input, available water, nutrients and colonization history. For national parks, this optimal state has been referred to by such terms as natural, naturally evolving, pristine and untouched. It implies that ecosystem structures and functions are unimpaired by human-caused stresses and that native species are present at viable population levels.

Approaches to Monitoring

Various possible approaches may be used in the development of a monitoring program. While there is overlap in these approaches, they have important theoretical differences. Briefly the four approaches are (1) reductionist, (2) integrated, (3) threat specific and (4) hypothesis testing. The reductionist approach of dissecting an ecosystem into its component parts (e.g., soil, water, insects, mammals) is rejected because, while it may be comprehensive, it is costly and cumbersome. Integrated monitoring aims to measure the so-called emergent properties of different levels in the recursive hierarchy of an ecosystem. It attempts to account for the differences in spatial and temporal scale and has been described as the preferred approach for ecosystem monitoring (Munn 1988). The problem with this approach is that it is inherently more difficult than the others and it is still in the development stages of application.

Despite these difficulties, for the purposes of this exercise, a monitoring scheme is developed for national parks that is two-pronged, based on both integrated and threat-specific monitoring approaches. The core of the system will be an integrated approach that we call ecosystem integrity monitoring, an approach which aims to understand changes at the ecosystem level. In addition, the specific stresses already identified for national parks in Canada, such as transboundary pollutants or visitor-wildlife interactions, require monitoring on a threat-specific or individual threat basis. Hypothesis testing is included as a tool for threat-specific monitoring. The incorporation of the above approaches into a monitoring scheme suitable to the national parks is shown below in Figure 1.

Considerations for a Monitoring Program

Key considerations in designing a monitoring program for ecosystem integrity include:

(1). Basic deficiencies exist in ecosystem science (Slobodkin 1988, Brown and Roughgarden 1990). For the most part we must be very humble in our management as we really do not know how ecosystems work.

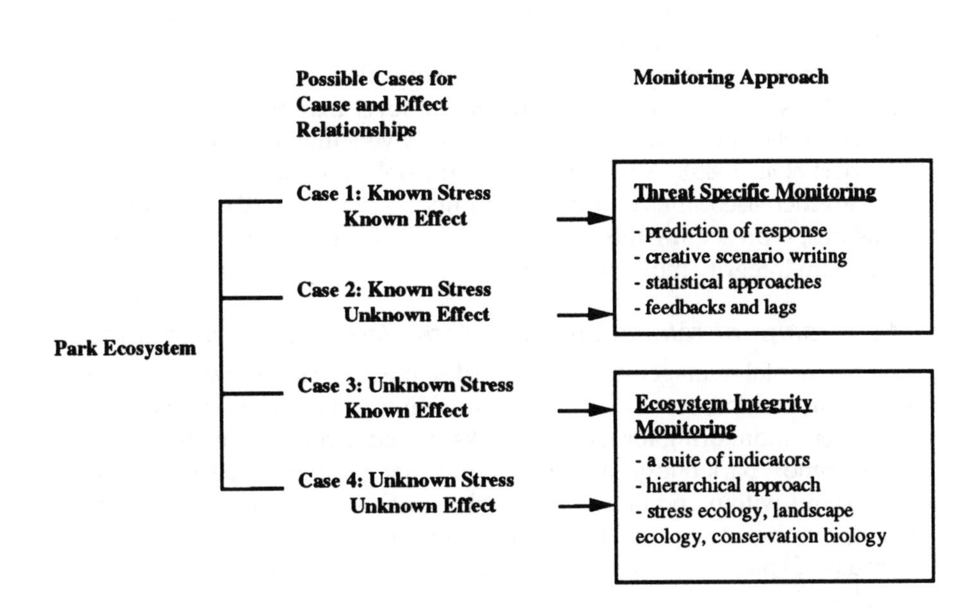

Figure 1: A two tiered framework for monitoring ecosystem integrity in Canadian national parks.

(2). Catastrophe and surprise characterize most ecosystems (Holling 1986, Kay 1991). Catastrophic changes are part of the normal functioning of ecosystems (e.g., spruce budworm outbreaks). Surprises occur when we are unaware of catastrophes or when ecosystems do not respond as we predict. Dealing with catastrophe and surprise must be integral to resource management.

(3). Stresses on ecosystems operate differently at various spatial and temporal scales (Sheehan 1984). A successful monitoring scheme must include measures from a variety of hierarchical levels, from individual to landscape. Individual level measures have the properties of early warning as individual organisms respond before the entire system. If changes are detected early, they are more reversible and thus more readily managed. However, changes at the individual level may not be reflected at higher levels as the system is assumed to have some resistance to stress (Rapport *et al.* 1985). Measures at higher levels of organization will provide an increased ability to trace the impacts of stress. Thus, a suite of measures, representing the range of hierarchical levels, is necessary for an ecosystem integrity monitoring program.

(4). A range of related social factors also must be monitored. Environmental management is an interdisciplinary science (Dorney, 1989) that requires input from many areas, including social and economic. Any ecosystem monitoring for national parks should include measures of visitation trends, park budget allocations, and changes in the type of activities carried out within and outside the park. These should not necessarily be exhaustive but designed to provide a comprehensive numerical picture. Because of space limitations, these factors are not discussed further in this chapter.

(5). Measures must be customized for specific ecosystems. This statement is considered self-evident in that one must deal with issues of context and scale.

Criteria for Monitoring Measures

Any measures used for monitoring the state of the ecosystems protected by national parks must be examined against some criteria. Several authors (Butler 1984, Miller 1984, Cook 1976, Schaeffer *et al.* 1988) have listed criteria that should be followed when developing diagnostic information on an ecosystem, or any system for that matter. A major problem in meeting these criteria is that there are very few data sets for ecosystems that give normative data or variances in the system. Decisions must nevertheless be made on whether any observed change in a system is part of a *natural* fluctuation or results from an external stress. Not all criteria can be followed in developing a strategy for monitoring in national parks. Some of the published criteria are very idealistic, such as the requirement that monitoring data must be indicative of the overall state of an ecosystem. No single measure has been developed that can satisfy that criteria. Others don't make sense in light of what is known about ecological science, such as the need to be comparable between different ecosystems. The differences between terrestrial and aquatic ecosystems are well established (Schlinder 1987, Gulati 1989).

I have taken the criteria that fit the goals of monitoring in national parks and added some criteria that are discussed in the *Considerations* section to develop the following new list of criteria to be used for national parks.

1. Monitoring measures should be relatively easily and reliably measured.
2. Monitoring measures should ideally have the capability to provide a continuous assessment from polluted to unpolluted (stressed to non-stressed) conditions.
3. Monitoring of ecosystem health should not depend solely on a single criterion, such as the presence, absence or condition of a single species. Any conclusions about ecosystem health should be based upon a suite of measures interpreted by experts.
4. Monitoring should reflect our knowledge of normal succession or expected sequential changes which occur naturally in ecosystems.
5. Measures used for monitoring should have a defined mean and variance whenever possible. If means and variances do not exist, data collection should be designed to establish them.
6. Ecosystem monitoring must be designed to accommodate the wide range of spatial and temporal scales, from individual to community to ecosystem.
7. Monitoring should be designed, wherever possible, to account for catastrophic changes that occur in ecosystems.
8. Monitoring must be based on the concept of ecosystems and not national park boundaries. National park boundaries in Canada were never designed to conform with ecosystem boundaries. The assessment of the state of ecosystems protected by national parks must be done on the basis of assessing the larger ecosystems of which national parks are a part.

9. Monitoring should be done in two ways — on the state of park ecosystems in general using the above criteria, as well as on specific threats known to exist (e.g., acid precipitation).
10. Monitoring measures must be designed for specific ecosystems because the sensitivity of various elements of ecosystem structure and function varies between ecosystems.
11. Monitoring should, whenever possible, provide for the early detection of change so that management action, if required, may be taken before the change becomes irreversible.
12. A monitoring program should be evaluated according to specific criteria and on a regular basis.

Threat-specific Monitoring in National Parks

Most of the current monitoring literature is devoted to threat-specific monitoring, such as the impact on ecosystems of heavy metals, ozone or human sewage. Lists of threats have been published for U.S. national parks (Machlis and Tichnell 1987) based on literature reviews and interviews with park staff. In Canada, many threats to national parks can be listed, but the mechanism for identification is somewhat random. The use of the general monitoring scheme for ecosystem integrity described in this report will provide one mechanism for identifying specific threats. But the monitoring scheme should still be supported by the existing network of staff knowledge, the Park Conservation Plan (part of the Canadian Parks Service's Natural Resource Management Process) and literature reviews. A scheme for monitoring specific threats is presented in Figure 2.

Selected Measures for Ecosystem Integrity Monitoring

Different schools of ecology, including stress response, ecology, conservation biology and landscape ecology were examined for measures of ecosystem integrity. The following measures were chosen because they met the criteria listed above.

Primary Productivity, the amount of organic matter produced by biological activity per unit area and time, has been shown to be susceptible to stress. (Schaeffer *et al.* 1988) Pine forests exposed to airborne pollutants invariably experience stunted needle growth and premature needle loss (Williams 1980, Mann *et al.* 1980). As production decreases, respiration often increases as energy is diverted to repair. Change in primary produc-

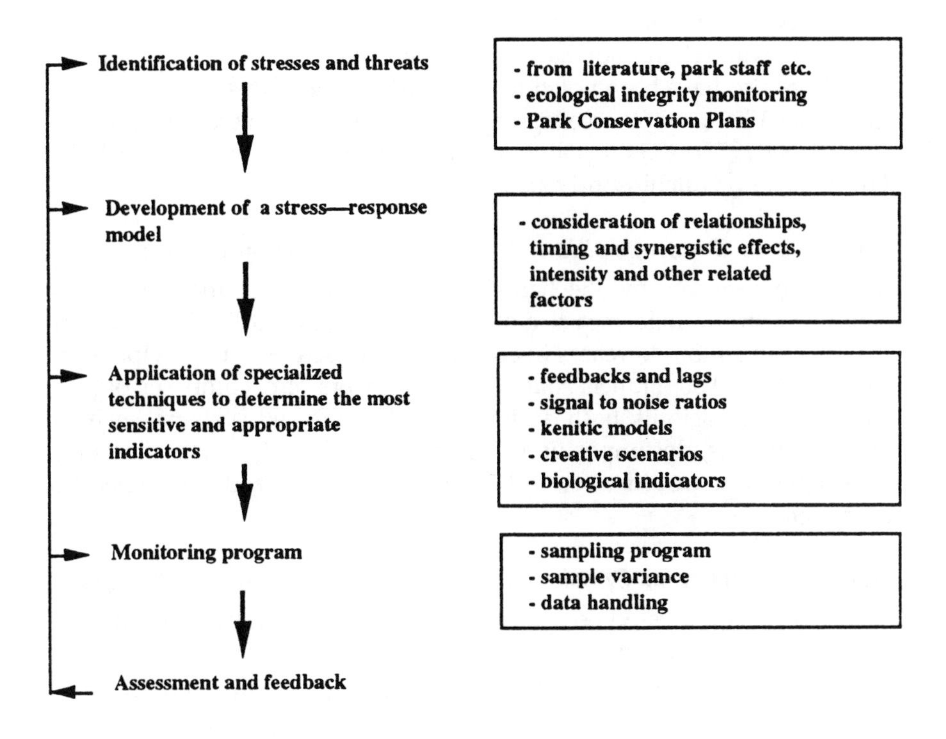

Figure 2: A schematic framework for the monitoring of specific threats in Canadian national parks.

tivity also can occur in the other direction in stressed systems such as aquatic systems, where phosphorus pollution had a fertilizing effect in Lake Erie (Regier and Hartman 1973). Measures of primary productivity are certainly possible, but traditional methods are time consuming and the sampling procedure is destructive. Recent advances have been made in measuring primary productivity by remote sensing and this may make such a measure of ecosystem health very useful.

Nutrient Cycling and Losses have been shown to respond predictably to various stresses (Odum 1985). At Hubbard Brook, New Hampshire, a deforested watershed had losses, via stream water runoff, of nitrate, calcium and potassium that were much larger than losses from an undisturbed reference watershed (Borman *et al.* 1974, Likens *et al.* 1978). Similar responses also have been noted for acid precipitation stressed ecosystems (Schindler 1987), whole tree logging in British Columbia (Kimmins 1977), fire (Weber 1977), and air pollutants in forests (Jackson and Watson 1977, Freedman and Hutchinson 1980a).

The Rate of Decomposition - The effects of stress on decomposition have been reviewed by Sheehan (1984a). He concluded that this factor could be a useful field measure of ecosystem health. It has been shown that decomposition decreases in ecosystems stressed by acid precipitation (Norten *et al.* 1980, Francis *et al.* 1980). Numbers of decomposer organisms and their activity both decline. In other stressed systems, such as clear-cut forests, decomposition rates rose significantly. As a measure, decomposition would have to be carefully normalized for a specific ecosystem.

Species Diversity or Species Richness has been shown to change for several kinds of stressed ecosystems, including SO_2 emissions on forests (Freedman and Hutchinson 1980a), radiation exposure on forests (Woodwell 1967), and pollutants on estuarine diatom communities (Patrick 1967). If change in species diversity is to be used as a measure of stress, then the choice of taxa and diversity indices must fit the ecosystem under consideration. Also, native diversity must be separated from non-native diversity.

Retrogression, or reversion to an earlier stage of succession, has been discussed by several authors as a trend exhibited by stressed ecosystems (Rapport *et al.* 1985, Schaeffer *et al.* 1988). Woodwell (1967) noted that pollutant stressors tend to cause changes that are just the reverse of those occurring during normal succession — the communities are simplified, niches are opened and succession begins again. The reversal of succession following stress has been reported in many areas, including terrestrial ecosystems following fire (MacLean *et al.* 1979) or aquatic systems following eutrophication (Francis *et al.* 1979).

Habitat Fragmentation has been called the greatest worldwide threat to forest wildlife (Rosenburg and Raphael 1986) and the primary cause of species extinction (Wilcox and Murphy 1985). The most studies on the effects on fragmentation have been made in temperate areas (Harris 1984, Forman and Godron 1986), with a concentration on forest birds. A very large multi-species study is currently underway in the tropics (Lovejoy *et al.* 1986). Fragmentation can be easily measured using the existing GIS technology. Fragmentation measures in parks should include a measure of human disturbance (e.g., the percentage of park developed or not in wilderness land use, length of developed edge in the park, the zone of human influence) as well as natural habitat fragmentation from disturbance elements such as fire and insect outbreaks. Fragmentation measures must be done on the larger ecosystem of which the park is only a part.

Minimum Viable Population size theory attempts to determine threshold levels for species survival over the long term (Soulé and Simberloff 1986). The susceptibility of a given species to extinction depends on many factors, the most important being body size, age at first reproduction, birth interval and susceptibility to catastrophe. Large bodied, long lived species such as redwood trees have a low rate of turnover (extinction and recolonization of a patch) and are more susceptible to extinction than small, short lived species like annual plants (Goodman 1987). Key species for monitoring should be chosen according to the defined categories, including top predators, rare species, and large body size organisms.

Minimum Area Requirements Perspectives from island biogeography (MacArthur and Wilson 1967), suggest that protected areas could not protect all species and that species persistence was directly related to the size of the protected area. Minimum area can be calculated for selected species, especially those with large territories (e.g., grizzly bears), rare species or species with sparse distribution. Calculation of minimum area must be made without regard to park boundaries.

Population Dynamics of Selected Species are required information to determine the minimum viable population size and population viability. Changes in population dynamics also will provide early warning of changes to the ecosystem. While it will not be feasible to measure all population parameters, an accurate measure of recruitment to the population and an estimate of total population size should be obtained. With some species it also will be desirable to obtain measures of age-specific mortality so life tables can be constructed. Practical difficulties can be reduced by careful selection of species. For example, great-horned owls are a major predator and easier to monitor than foxes. Similarly, plants such as large trees can be used to monitor large body size organisms. These measures are summarized by their position in the hierarchical scale in Figure 3.

Interpreting the Data

The approach to developing a monitoring program outlined in this chapter might be criticized because monitoring measures were chosen before there was a discussion of the interpretation of those measures. In contrast, the USEPA'S Environmental Monitoring and Assessment Program (EMAP) is driven by the prior development of assessment goals. I contend that the approach outlined in this paper is also driven by assessment goals, the difference being that the goals are defined in more general terms. All the indicators developed for EMAP are linked to detailed assessment endpoints (Charles *et al.* 1990), based on determining the exact intensity and extent of a particular stress. This approach is too restrictive for the development of a monitoring program in national parks. The kinds of assessment and our ability to assess depend upon the state of ecosystem science, as well as, prevailing attitudes of the day. Any long term monitoring program must allow for flexibility, if it is going to be useful in the long term and not become quickly dated. Thus, the approach used in this chapter is to let the monitoring framework be driven by the more general goals of measuring ecological integrity and health. The approach is to monitor for both *known and unknown* stresses. The definite endpoint strategy used in EMAP is limited by the use of only *known* stresses. This is short sighted because new stresses will undoubtedly be defined and many of the known stresses act together to produce combined and unpredictable effects.

The critical question remaining is of how to interpret the data that is monitored. There are two general approaches that might be followed. The first is the use of reference systems that compare the state of a test ecosystem against a known or reference ecosystem. This has been done

successfully by Karr (1981) for freshwater streams in the midwest United States. The health of the stream is assessed by comparison with index values from healthy, local reference streams. For national parks, the notion of reference ecosystems is not applicable. National parks are generally large and unique, and are often the least stressed ecosystems within the natural region they represent. Comparison outside the natural region is probably not relevant. However, the principle of comparison to reference standards is a good one and will have many general applications. For example, there are many published examples of decomposition rates for different ecosystems. These values could be used to compare values obtained from monitoring.

A second method of interpreting monitoring information is to analyze trends. This method is recommended for parks and, in fact, this monitoring program is based on providing trend analyses. The monitoring measures all are chosen to show trends that would be expected in stressed ecosystems. The inability to retain nutrients, loss of native species diversity, changes in community diversity and habitat fragmentation are all examples of trends associated with stressed ecosystems. They must be assessed over time as trend, interpreted by knowledgeable people, preferably by a group of knowledgeable people.

The premise of this monitoring scheme is that ecosystems are nested or recursive hierarchies and that a suite of measures is required to monitor different levels in the hierarchies. Some measures are useful as early warning indicators (e.g., individual growth rates). Others are sensitive indicators of a particular stress and provide unambiguous interpretation of cause and effect (e.g., accumulator species such as clams). Still others indicate trends in ecological processes (e.g., decomposition rates) or system level changes (e.g., habitat fragmentation). In the absence of an ecosystem theory that explicitly shows how different hierarchical scales interact, we are left to define ecosystems, their exposures to stress and subsequent responses, operationally (Kelly and Harwell 1990).

In addition to ecological measures, it is strongly recommended that some relevant social and economic measures also be assessed. These could be drawn up to be relevant to each specific park. These measures may include such things as number of dollars and person-years spent on resource management, and the amount of progress toward the objectives for cooperative agreements between the park and external land-use agencies. These types of measures are best made against defined goals for ecosystem management (Agee and Johnson 1988). Such goal setting must come from park management plans if parks are to conduct ecosystem-based management.

The utility of trend analysis will be revealed when this monitoring scheme is field tested. There are many ways to present the results of the trend analysis. An index or indices could be developed, such as the total number of negative trends from one reporting period to the next. Such cumulative indices are used frequently in society to present complex information. The Consumer Price Index, the Dow-Jones Industrial Average and the Gross National Product are all presentation indices that simplify very complex economic systems for the purposes of reporting. Such indices are valuable in that they synthesize a great deal of information. Their drawback is that no single numerical index can be expected to represent the status of a complex ecosystem. The economic indicators noted above are notoriously weak descriptors of economic systems. The complexity of economic systems pales by comparison to ecosystems. Therefore, interpretation might be best performed by that great integrator, the human brain.

One successful method of presenting complex ecological data has been developed by Nip *et al.* (1990) working in Holland. It is called the "amoeba" approach and it graphically presents a series of environmental measures without resorting to an index. The integration is done visually by the observer of the "amoeba" diagram. An example amoeba diagram from Nip *et al.* (1990) is shown in Figure 4. This method of reporting is based on developing policy goals or yardsticks for each of the chosen measures. An example might be to keep non-native species of flora below 10% of the total flora. That measure of the percentage of non-native flora becomes one segment in the pie of the amoeba diagram and the yardstick of 10% becomes the circumference of the circle. The number of measures can be increased or decreased by simply dividing the circle into more or fewer slices. The circumference of the circle always refers to the policy yardstick. Thus if all measures were as desired, the amoeba diagram would look like a circle with perfectly equal pie shaped wedges. Deviations from the yardstick extend the pie segment representing the particular measure away from the circle circumference. Thus, the more "amoeba-like" the diagram becomes, the worse the environmental situation. Such a tool for presenting data will not do the job of analyzing the ecological status of the park. However, it does provide a graphic method to summarize a lot of complex information and it can combine a number of very different measures, including ecological, social and economic.

Conclusions

Monitoring is an integral part of the management of natural systems and is essential if the Canadian Parks Service is to meet its mandate of ensuring ecological integrity as stipulated in the National Parks Act (1989). Monitoring for ecosystem integrity is not a substitute for the traditional resource surveys done by the Canadian Parks Service, nor does it replace

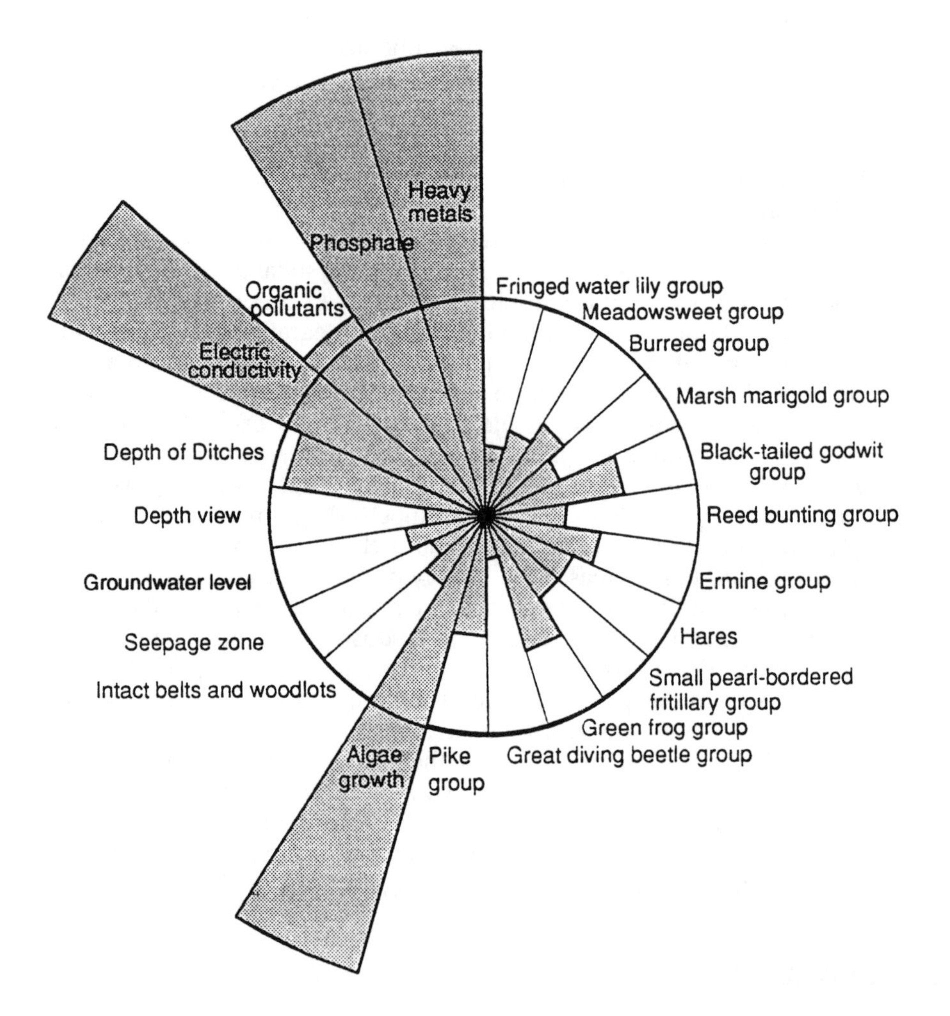

Figure 4: An example of an Ameoba diagram (from Nip *et al.* 1990).

the existing Natural Resource Management Planning Process. It should be seen as a feedback system for park management. A monitoring database will allow managers to track both the state of the park's environment and specific threats to the park. A monitoring database also could easily form the basis for the required "State of the Parks Reports" and have tremendous value for "State of the Environment" reporting.

Monitoring in national parks must be accomplished in two distinct ways if it is to achieve all its goals. It must first monitor the general state of ecosystem health or integrity. This has been done by using measures from several schools of ecology, representing different hierarchical levels (Figure 3). The second approach that is required is that of monitoring specific threats or stresses.

The Canadian Parks Service can make a fundamental and significant contribution to environmental management in Canada. With 34 parks in diverse natural regions across Canada, from the high arctic to the Carolinian zone, a ready-made system of potential *environmental monitoring sites* are available that could be of inestimable value in assessing the environmental health of Canada.

REFERENCES

Agee, James K. and Darryll R. Johnson. 1988. A direction for ecosystem management. In: Johnson, Darryll R. and Agee, James K. (eds.) *Ecosystem Management for Parks and Wilderness*. University of Washington Press.

Borman, F.H., G.E. Likens, T.G. Siccama, R.S. Pierce, and J.S. Eaton. 1974. The export of nutrients and recovery of stable conditions following deforestation at Hubbard Brook. *Ecological Monographs*. 44, 225-277.

Brown, J.H. and J. Roughgarden. 1990. Ecology for a changing earth. *Bull. Ecol. Soc. America*. 71:173-188.

Bruntland, G. 1887. *Our Common Future*. New York: Oxford University Press. 400 pp.

Butler, G.C. 1984. Introduction. In: Sheehan, P.J., Miller, D.R., Butler, G.C. and Bourdeau, D.R. (eds.) *Effects of Pollution at the Ecosystem Level*. pp 1-14. 1984 SCOPE. John Wiley and Sons.

Cairns, J. 1977. Quantification of biological integrity. In: Ballentine, R.K. and Guarraia, L.J. (eds.). *The integrity of water.* U.S. E.P.A. Office of Water and Hazardous Materials. U.S. Govt. Printing Office, Washington, D.C.

Charles, Donald F., Joan P. Baker, Jeffrey M. Klopatek, Charles M. Knapp, David R. Marmorek, and Kent W. Thorton. 1990. A research strategy to develop ecological indicators for the environmental monitoring and assessment program (EMAP). Paper presented at the Int. Symp. on Ecol. Indicators. Fort Lauderdale, Florida, October, 1990.

Cook, S.E.K. 1976. Quest for an index of community structure sensitive to water pollution. *Environ. Pollut.*, 1, 57-71.

Dorney, Robert S. 1989. *The Professional Practice of Environmental Management.* Dorney, Lindsay (ed.). Springer-Verlag. 228 pp.

Forman, R.T. and M.G. Godron. 1986. *Landscape Ecology.* John Wiley and Sons, New York. 619 pp.

Francis, A.J., H.L. Quinby, and G.R. Hendry. 1980. Effect of lake pH on microbial decomposition of allochthonous litter. In: Hendry, G.R. and Teasley, J. (eds.), *Acid Rain: Aquatic Effects, ACS Acid Rain Symposium Proceedings.* Ann Arbor Science Publishers, Ann Arbor, MI.

Francis, George R., A.P. Lino Grima, Henry A. Regier, and Thomas H. Whillans. 1985. *A Prospectus for the Management of the Long Point Ecosystem.* Great Lakes Fishery Commission Tech. Rep. No. 43. Ann Arbor, Mich.

Francis, George R., John J. Magnuson, Henry A. Regier, and Talhelm, R. Daniel. 1979. *Rehabilitating the Great Lakes Ecosystems.* Great Lakes Fishery Commission Tech. Rep. No. 37. Ann Arbor, Mich.

Freedman, B. and T.C. Hutchinson. 1980). Long term effects of smelter pollution at Sudbury, Ontario on surrounding forest communities. *Can. J. Bot.* 58: 2123-40.

Freedman, B. and T.C. Hutchinson. 1980b. Smelter pollution near Sudbury, Ont., Canada and effects on forest litter decomposition. In: Hutchinson, T.C. and Havas, M. (eds.) *Effects of Acid Precipitation on Terrestrial Ecosystems.* Plenum Press, New York.

Goodman, D. 1987. The demography of chance extinction. In: Soulé, M.E. (ed.) *Viable Populations for Conservation*. Cambridge University Press. pp 11-35.

Gulati, R.D. 1989. "Concept of ecosystem stress and its assessment," In: *Ecological Assessment of Environmental Degradation, Pollution, and Recovery*. Elsevier, Amsterdam. pp. 55-80.

Harris, L.D. 1984. *The Fragmented Forest: Island Biogeography Theory and the Preservation of Biotic Diversity*. Univ. of Chicago Press. 211 pp.

Holling, C.S. 1986. The resilience of terrestrial ecosystems: local surprise and global change. In: Clark, W.C. and Munn. R.E. (eds.), *Sustainable Development of the Biosphere*. Cambridge University Press, Cambridge. pp. 292-317.

Jackson, D.R. and A.P. Watson. 1977. Disruption of nutrient pools and transport of heavy metals in a forested watershed near a lead smelter. *J. Environ. Qual.*, 6:331-38.

Karr, James R. and D.R. Dudley. 1981. Ecological perspective on water quality goals. *Environmental Management*, 5:55-68.

Kay, James R. 1989. A Non-Equilibrium Thermodynamic Framework for Discussing Ecosystem Integrity. For Int.Joint Comm. and Great Lakes Fisheries Comm. Workshop on Ecosystem Integrity in the Context of Surprise. Burlington, June 1988. 28 pp.

Kelly, J.R. and M.A. Harwell. 1990. Indicators of ecosystem recovery. *Environmental Management*. 14(5):527-545.

Kimmins, J.P., 1977. Evaluation of the consequences for future tree productivity on the loss of nutrients in whole tree harvesting. *For. Ecol. Manage.* 1:169-183.

Leopold, Aldo. 1939. A biotic view of land. *Journal of Forestry 37*. (September) pp. 727-730.

Likens, G.E., F.H. Bormann, R.S. Pierce, and W.A. Reiners. 1978. Recovery of a deforested ecosystem. *Science*. 199: 492-496.

Lovejoy, T.E., R.O. Jr. Bierregaard, A.B. Rylands, J.R. Malcom, C.E. Quentella, L.H. Harper, K.S. Jr. Brown, A.H. Powell, G.V.N. Powell, H.O.R. Schubart, and M.B. Hays. 1986. Edge and other effects of isolation on Amazon forest fragments. In: Soulé, M.E. *Conservation Biology. The Science of Scarcity and Diversity*. Sinaurer Ass., Inc. Sunderland, Mass. pp. 257-285.

MacArthur, Robert H. and Edward O. Wilson. 1967. *The Theory of Island Biogeography*. Princeton University Press. 203 pp.

Machlis, G.E. and D.L. Tichnell. 1987. Economic development and threats to national parks: a preliminary analysis. *Environ. Conserv.* 14 (2): 151-56.

MacLean, D. A., S.J. Woodley, M.G. Weber, and R.W. Wein. 1979. Fire and Nutrient Cycling. In: Wein, Ross W. and MacLean, David A. (eds.). *The Role of Fire in Northern Circumpolar Ecosystems*. SCOPE 18. John Wiley and Sons. pp. 111-127.

Mann, L.K., S.B. McLaughlin, and D.S. Schriner. 1980. Seasonal physiological responses of white pine under chronic air pollution stress. *Environ. Exp. Bot. 20*:99-105.

Miller, Donald R. 1984. Chemicals in the environment. In: Sheehan, P.J., Miller, D.R., Butler, G.C. and Bourdeau, D.R. (eds.) *Effects of Pollution at the Ecosystem Level*. 1984 SCOPE. John Wiley and Sons. pp. 7-14.

Munn, R.E. 1988. The design of integrated monitoring systems to provide early indications of environmental/ecological changes. *Environ. Mon. & Assess.* 11: 203-17.

National Parks Act. 1988. Parliament of Canada, Ottawa, Canada.

Nip, M.I., J. Latour, F. Klijn, P. Koster, C.L.G. Groen, H.A. Udo de Haes, and H.A.M. de Kruijff. 1990. Environmental quality assessment of ecodistricts: A comprehensive method for environmental policy. Symposium on Ecological Indicators, Ft. Lauderdale, Florida.

Norten, S.A., D.W. Hansen, and R.J. Compana. 1980. The impact of acid precipitation and heavy metals on soils in relation to forest ecosystems. In:*Effects of Air Pollutants on Mediterranean and Temperate Forest Ecosystems*. U.S.D.A., Forest Service Tech. Rep. PSW-43, Berkley, Calif. pp 152-164.

Odum, E. P. 1985. Trends expected in stressed ecosystems. *Bioscience*. 35(7): 419-422.

Patrick, R. 1967. Diatom Communities in Estuaries. *Am. Assoc. Adv. Sci. Publ.* 83: 311-315.

Rapport, D.J., H.A. Regier, and T.C. Hutchinson. 1985. Ecosystem behavior under stress. *Am. Nat.* 125: 617-640.

Regier, H.A. and W.L. Hartman. 1973. Lake Erie's fish community: 150 years of cultural stress. *Science*. 180:1248-1255.

Rosenburg, K.V. andM.G. Raphael. 1986. Effects of forest fragmentation on vertebrates in Douglas-fir forests. In: Verner, J., Morrison, M.L., and Ralph, C.J. (eds.) *Wildlife 2000. Modelling Habitat Relationships of Terrestrial Vertebrates*. University of Wisconsin Press. pp. 327-9.

Schaeffer, David J., Edwin E. Herricks, and Harold W. Kerster. 1988. Ecosystem health: I. Measuring ecosystem health. *Environ. Management*. 12(4): 445-55.

Schindler, J.E. 1987. Detecting ecosystem responses to anthropogenic stress. *Can. J. Fish. Aquat. Sci.* 44(Supp.1):6-25.

Sheehan, Patrick J. 1984a. Effects on community and ecosystem structure and function. In: Sheehan, P.J., Miller, D.R., Butler, G.C. and Bourdeau, D.R. (eds.). *Effects of Pollution at the Ecosystem Level*. 1984 SCOPE. John Wiley and Sons. pp. 51-100.

Sheehan, Patrick J. 1984b. Functional Changes in the Ecosystem. In: Sheehan, P.J., Miller, D.R., Butler, G.C. and Bourdeau,D.R. (eds.). *Effects of Pollution at the Ecosystem Level*. 1984 SCOPE. John Wiley and Sons. pp. 101-145.

Slobodkin, L.B. 1988. Intellectual problems of applied ecology. *Bioscience.* 38 (5): 337-42.

Soulé, Michael E. and D. Simberloff. 1986. What do genetics and ecology tell us about the design of nature reserves? *Biological Conservation.* 35:19-40.

Weber, M.G. 1977. Nutrient Redistribution Following Fire in Tundra and Forest-Tundra. MSc. Thesis, Univ. New Brunswick, Fredericton, N.B. 69 pp.

Wilcox, B.A. and D.D. Murphy. 1985. Conservation strategy: The effects of fragmentation on extinction. *American Naturalist.* 125:879-887.

Williams, W.T. 1980. Air pollution disease in the Californian forests: a baseline for smog disease on ponderosa and Jeffery Pines in the Sequoia and Los Padres National Forests, California. *Environ. Sci. Technol.* 14:179-182.

Woodwell, G.M. 1967. Effects of ionizing radiation on terrestrial ecosystems. *Science.* 138, pp. 572-7.

An Approach to the Development of Biological Sediment Guidelines

TREFOR B. REYNOLDSON and MICHAEL A. ZARULL
Lakes Research Branch, National Water Research Institute
CCIW, Burlington, Ontario

Introduction

Sediments contaminated with nutrients, metals, organics and oxygen demanding substances can be found in freshwater and marine systems throughout the world. While some of these contaminants occur in elevated concentrations as a result of natural processes, the presence of many is a result of human activity. Aquatic sediments with elevated levels of contaminants can be found in any low energy area that is the recipient of water associated with urban, industrial or agricultural activity. Such low energy depositional zones are found in many nearshore embayments, and lake and river mouth areas. Sediments are usually designated as contaminated, almost exclusively, on the basis of bulk chemical concentrations by those agencies responsible for the protection and management of the environment (USEPA 1977, Ontario Min. Environment 1988). However, the measurement of chemical concentrations in the sediments does not address whether the contaminants present are exerting biological stress and whether this stress will continue after the primary sources (point and nonpoint) of pollution have been controlled.

Efforts to measure and mitigate the impacts of sediment associated contaminants arose from the continuing need to remove and dispose of large volumes of material for navigational purposes. Initially, dredge spoils were dumped offshore; however, concern over the impacts of such practices led to confining the more contaminated material. The need to quickly and inexpensively assess the potential environmental impacts of dredged material, prior to disposal, resulted in the development of chemical specific criteria. Continued concern over the degree of environmental protection offered by these chemical guidelines, the lack of uniform, international criteria and the introduction of plans to remediate degraded areas in the Laurentian Great Lakes have prompted the development of more comprehensive sediment assessment approaches and corresponding evaluation criteria. Many of these proposals have strongly supported the need for a series of bioassessment techniques, along with appropriate criteria or objectives, which identify the type and severity of stress being exerted, as well as contaminant bioavailability (Dillon and Gibson 1986, IJC 1988a, Reynoldson and Zarull 1989).

Assessing Contaminated Sediments

A number of approaches have been used to assess sediment contamination; however, they all employ either chemical or biological measures, or in some cases, both. Once the chemical concentration, biological response or state has been ascertained, the results are compared to some predetermined objective, criteria or standard. Chemical analyses are most frequently performed on the bulk sediment for total compound concentration (e.g., total PCBs). Analyses are done less often on pore water. Also, chemical species or congeners are measured infrequently in either bulk sediments or pore water. These chemical concentrations are then compared to background concentrations (pre-historic), existing water quality criteria, or to known biological effects levels. The latter method includes the equilibrium partitioning approach, the screening level concentration approach and the apparent effects threshold (Beak 1987, Persaud *et al.* 1989, USEPA 1989). While each of these approaches to developing sediment objectives or criteria have their own advantages and disadvantages, they collectively suffer from the need to use inference of biological impact rather than demonstrating effects. These approaches should provide a high degree of protection to the aquatic ecosystem; however, they are unlikely to prove adequate in justifying expensive removal, treatment or mitigation of in-place pollutants as part of a comprehensive remedial action plan for nearshore areas.

Biological assessments employ, either singly or in combination, structural and functional measures of impact. Structural assessments examine changes in community composition and/or changes to indicator species and relate these to anthropogenically induced stresses. While the examination of ecosystem structural integrity should ideally include all biotic components of the ecosystem, such detail is prohibitively expensive and time consuming and not necessarily required (IJC 1987). Instead, the most appropriate biotic community component is assessed. In the case of sediment work, the most appropriate group of organisms is the benthic invertebrates. These organisms reside in or on the sediments; they are sessile; their populations are relatively stable in time, with life cycles of one or more years; their taxonomy is reasonably well established at the genus, and in some cases the species level; and their responses to environmental changes have been studied extensively.

Functional, biological assessment can be broadly defined as the measurement of processes or ecosystem responses, and can use both taxonomic and nontaxonomic parameters (Mathews *et al.* 1982). Taxonomic tests include measures of species colonization or emigration rates and the rate of re-establishment of equilibrium densities following perturbation (Cairns *et al.* 1979). The more frequently used nontaxonomic tests include acute (short term) and chronic (longer term) measures of behaviour, biochemical/physiological changes, bioaccumulation, genotoxicity (mutagenesis, carcinogenesis, teratogenesis), reproductive failure and death. While many of these tests can be conducted *in situ*, the need for strict control over exposure conditions means that most bioassays are conducted in the laboratory. In addition, most of these tests are conducted using a single species which may not be indigenous to the area under investigation. If a toxicity assay approach is to be used, Adams and Darby (1980) outline a number of decisions or questions which must addressed:

1. What bioassay organism should be used?
2. What positive or negative controls are used?
3. Which endpoints are measured?
4. How are relationships between endpoints and assays determined and how are relative sensitivities compared?
5. What are the minimum number of assays required in a battery of tests?
6. What are appropriate test conditions?

Although functional tests can provide quantifiable relationships between contaminants and organisms, their lack of environmental realism and possibly, their inability to incorporate ecosystem complexity limits their investigative utility and capability to establish actual impairment in the environment. The obvious solution to this dilemma is to use both structural and functional methods to examine stress.

Recent initiatives and reviews in the United States and Canada (IJC 1987, 1988a, Beak 1987, Persaud *et al.* 1989, Giesy and Hoke 1990, USEPA 1989) and the Netherlands (van de Guchte and van Leeuwen 1988) describe current thinking on methods of developing criteria and/or objectives. Several methods are now under consideration, all of which provide chemically based sediment criteria (e.g., Persaud *et al.* 1989, Min. Transport & Public Works 1989). However, since the fundamental reason for sediment objectives is to protect a sustainable and reproducing aquatic biota, a more direct approach is required.

An Ecological Approach to Developing Sediment Guidelines

The existence of a sustainable and reproducing biota is determined by the occurrence, through time, of an assemblage of species that is self-maintaining and resilient to environmental fluctuations. Such communities can be defined through the collection of physical, chemical and biological data; however, the primary purpose for the collection of physio-chemical data is the interpretation and identification of the causative factors that result in the observed structure of the biological community. It seems therefore, logical to us, to define the existence of sustainable and self-reproducing aquatic fauna by using biological guidelines or objectives. The most promising approaches in determining impact from, and causative factors in, contaminated sediments are measurements of benthic invertebrate community structure and sediment toxicity. We would suggest the two components be used in the definition of biological sediment guidelines, sediment toxicity and quantification of *in situ* effects.

Other authors have made similar proposals for assessing and managing contaminated sediments. Chapman and co-workers (Chapman and Long 1983, Long and Chapman 1985, Chapman 1986) have described the sediment triad approach, which incorporates aspects of sediment chemistry, toxicity testing and community structure analysis. The International Joint Commission has suggested a sediment management strategy that incorporates both assessment and remediation (IJC 1987, 1988a, 1988b). The United States is embarking on a program to examine various approaches to sediment assessment and remediation and Canada is also addressing these issues via its Great Lakes Action Plan. However, the major objections still posed to employing benthic community structure analysis for setting sediment guidelines is their lack of universality (i.e., they are completely site-specific) and the inability of researchers to establish quantitative objectives for their application (i.e., what should the community `look' like?). We feel that by the very nature and complexity of sediment-contaminant-biota relationships and the desire to apply sediment guidelines to initiate large scale, costly remedial action in the environment, universal guidelines and/or objectives are not possible. Methods which address the latter criticism are proposed in this chapter.

Recent developments in the application of multivariate analysis show extremely promising results in interpreting changes in community structure. There appears to be a general agreement among European authors (Metcalfe 1989) on the need to define reference communities based on chemical, physical, geological and geographical features, unrelated to pollution, as a first step toward identification of the best achievable communities for each type of water. The greatest effort in this direction has been made in the United Kingdom (Wright *et al.* 1984, Armitage *et al.* 1987, Moss *et al.* 1987), but similar studies have been conducted in North America (Corkum and Currie 1987), continental Europe (Johnson and Wielderholm 1989) and South Africa (King 1981). As a result of an extensive data collection program at nonpolluted sites in the U.K. (Wright *et al.* 1984), a number of natural communities were identified. Using selected and easily measured environmental variables, such as latitude, substrate type, temperature and depth, it was possible to predict correctly the community occurring at more than 75% of 268 sites, and at more than 90% of the sites the predicted community was identified to the correct or the next most similar community. At lower levels of community detail, even greater accuracy of prediction was observed. Other studies have shown a similar predictive capacity (Table 1), the accuracy of prediction ranging from 68% - 90%, this in habitats ranging from large lakes to small streams on three continents. In other words, based on a few physical variables the expected community assemblage at a location can be defined from predictive equations. Such predicted communities or key species in the communities could form site specific guidelines, and could be compared against the observed species composition, thus determining whether the guideline is being met. While, to date, this approach has only been used on community assemblages defined by species as classifying variables, there is no reason why the same approach cannot use functional variables, such as bioassay endpoints, to classify predictable toxicity response categories. To illustrate the type of criteria such an approach would provide, data from the Detroit River (Reynoldson and Zarull 1989) have been used.

Detroit River

In 1980, a sediment survey of the Detroit River produced a data matrix that included information from 59 stations for both benthic invertebrate community structure and sediment physics and chemistry (Thornley and Hamdy 1984). The proposed approach for developing biological guidelines has four components: classification of reference communities, development of a model to predict the biological community from sediment characteristics, the application of the model to test sites, and determination of impairment by comparison of existing species with predicted species, using a set of numerical guidelines.

Table 1: Performance of classification and discrimination analysis in predicting benthic invertebrate community assemblages from physico-chemical data.

HABITAT	SITES	COMMUNITIES	PREDICTOR VARIABLES	% CORRECT	AUTHOR
RIVERS	268	16	28	76.1	Write *et al.* 1974
RIVERS	268	8	28	79.1	Moss *et al.* 1974
STREAMS	79	5	26	70.9	Corkum & Currie 1987
RIVERS	45	6	5	68.9	Ormerod & Edwards 1987
RIVERS	54	5	9	79.6	King 1981
LAKES	68	7	11	90.0	Johnson & Wielderholm 1989
RIVERS	29	3	4	83.0	Reynoldson & Zarull this volume

Classification of Biological Communities in Reference Sites

To illustrate the process of defining reference communities the 29 stations on the Canadian side of the Detroit River were used (Figure 1), as these stations are known to represent a "clean" community (Reynoldson and Zarull 1989). Community assemblages were classified using average linkage clustering, K-Means, a divisive clustering technique (Wilkinson 1990), and ordination. Twenty benthic taxa were used as grouping variables. The average linkage clustering indicated three to six clusters were worthy of investigation. Using K-Means analysis, a number of communities (clusters) from three to eight were examined. These are presented in PCA ordination space on the first three components (Figure 2) which explained 79% of the variance. The three cluster solution was deemed most appropriate for use in defining reference communities as several of the clusters produced beyond this were formed from single sites. The cluster boundaries have been shown on the ordination plot (Figure 2a b) for the first three principal components, and show clusters 4, 5 and 6 to be comprised of single sites and that clusters 7 and 8 are located within the boundaries of cluster 1, therefore, the likelihood of discriminating these communities is small. The F ratios and component loadings show five taxa contributed significantly to differentiating between these communities (Table 2) and are therefore likely to be useful as numerical criteria in representing the three community assemblages.

Prediction of Expected Assemblages

An overriding problem in the application of biological guidelines, as opposed to a chemical approach, is the much greater site specificity requiring a series of biological values. In the Detroit River illustration, three sets of guidelines would be developed based on the three community assemblages. Consequently, a technique is required for selecting the most appropriate set of guidelines for any given site, based upon the natural local conditions of the environment. This requires a method that allows the prediction of the type of biological community that is expected at a site. Discriminant analysis is a statistical technique that permits groups to be distinguished using a set of discriminating variables on which the groups are expected to differ. In the case of benthic invertebrate community structure, the groups are determined by the community assemblages as identified by clusters formed in cluster analysis. The

Table 2: Numbers of various taxa and their relative contribution in three community assemblages from reference sites in the Detroit River.

TAXA (No. m^2)	F Ratio (K means)	Cluster 1 (n=23)		Cluster 2 (n=3)		Cluster 3 (n=2)	P.C. 1	P.C. 2	P.C. 3
		Mean	S.D.	Mean	S.D.	Avg.			
Ephemeroptera	5.5	126	(99)	304	(193)	582	0.832	0.420	-0.079
Tubificidae	144.5	385	(409)	5523	(843)	2546	0.687	-0.561	0.220
Isopoda	10.2	1	(2)	0	(0)	23	0.686	0.465	-0.477
Hirudinoidea	5.0	1	(2)	7	(5)	4	0.626	-0.455	-0.271
Odonata	22.3	0	(0)	5	(3)	0	0.466	-0.687	0.247
Amphipoda	0.9	44	(64)	7	(5)	88	0.373	0.659	0.247
Chironomidae	0.1	270	(448)	338	(232)	141	0.243	0.312	0.839

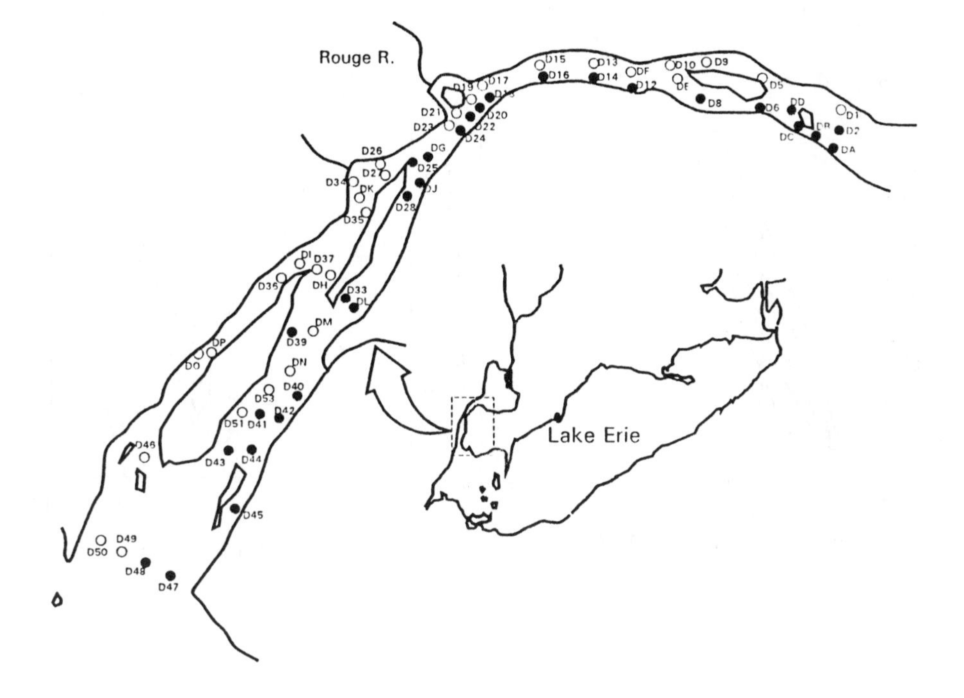

Figure 1: Sample site locations in the Detroit River (open circles - test sites, closed circles - reference sites).

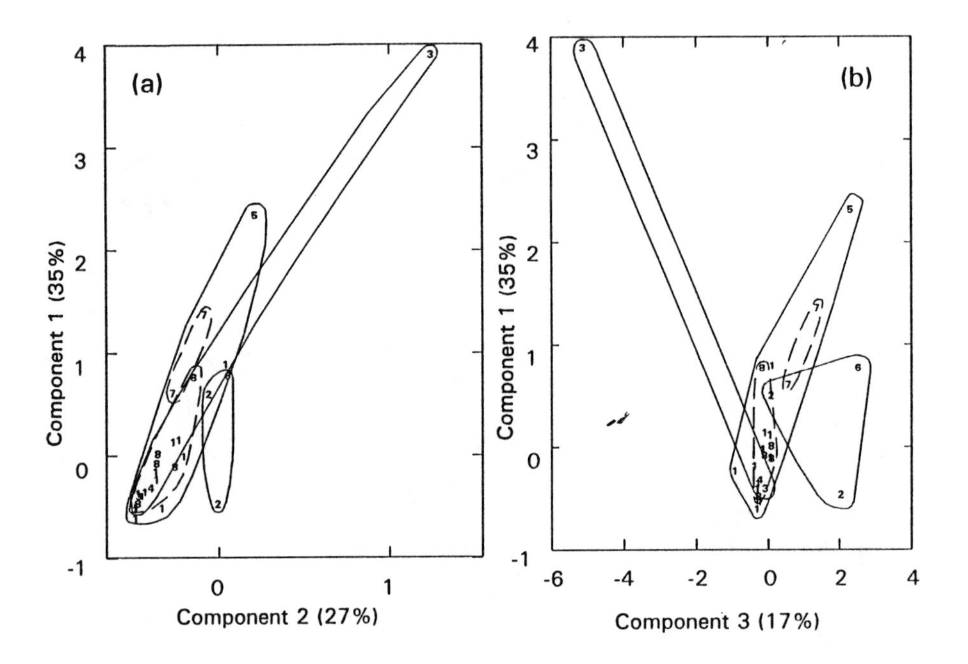

Figure 2: Ordination plots (PCA) of reference sites showing cluster boundaries (solid lines - 3 cluster solution, dashed lines - 8 clusters), on first three components (numbers indicate cluster associations).

discriminating variables will be values that measure differences in the physical and chemical characteristics of the sediment that are likely to result in different species being present. For the purposes of sediment guidelines, it is important that the discriminating variables are minimally impacted by pollution, as it is necessary to determine the community that is most likely to be present if the site were clean. In this illustrative case, four potentially important discriminating variables were used: the percent of sand, silt and clay, and the percent loss on ignition (LOI). The first three variables describe the physical form of the sediment. For example, a silty habitat could be expected to be dominated by burrowing organisms such as tubificid oligochaetes and chironomids whereas a sandy or rocky substrate is more likely to be dominated by amphipods and Ephemeroptera. The fourth discriminating variable, LOI, is an estimate of the amount of organic carbon present in the sediment which is related to food availability and thus may relate to the abundance of organisms. To confirm that the use of three assemblages was appropriate, multiple discriminant analysis was also conducted using four and eight groups from the classification analysis (Figure 3). The discriminant statistics indicated non-significance ($P > 0.05$) for differences between group centroids (Table 3) for both four and eight clusters, further confirming the validity of three community assemblages. Furthermore, the percentage of reference sites correctly predicted using three clusters was as high as 83%, but declined to 59% and 35% respectively using four and eight clusters. The actual predicted cluster membership for the three clusters shows that the four variables are good discriminators for the reference sites (Table 3).

Predicting Test Sites

The third component, in the development and use of biological guidelines, is the application of the model to the test sites (Figure 1) and a comparison of the predicted community with the actual species present. Application of the multiple discriminant model to the test sites resulted in the predicted communities shown in Figure 3. Of the 30 test sites, 22 were predicted as having community 1, four as community 2 and four as community 3. To determine whether sites are impacted, a method other than simple comparison, is required to compare the observed community with that predicted by multiple discriminant analysis from the reference set of sites.

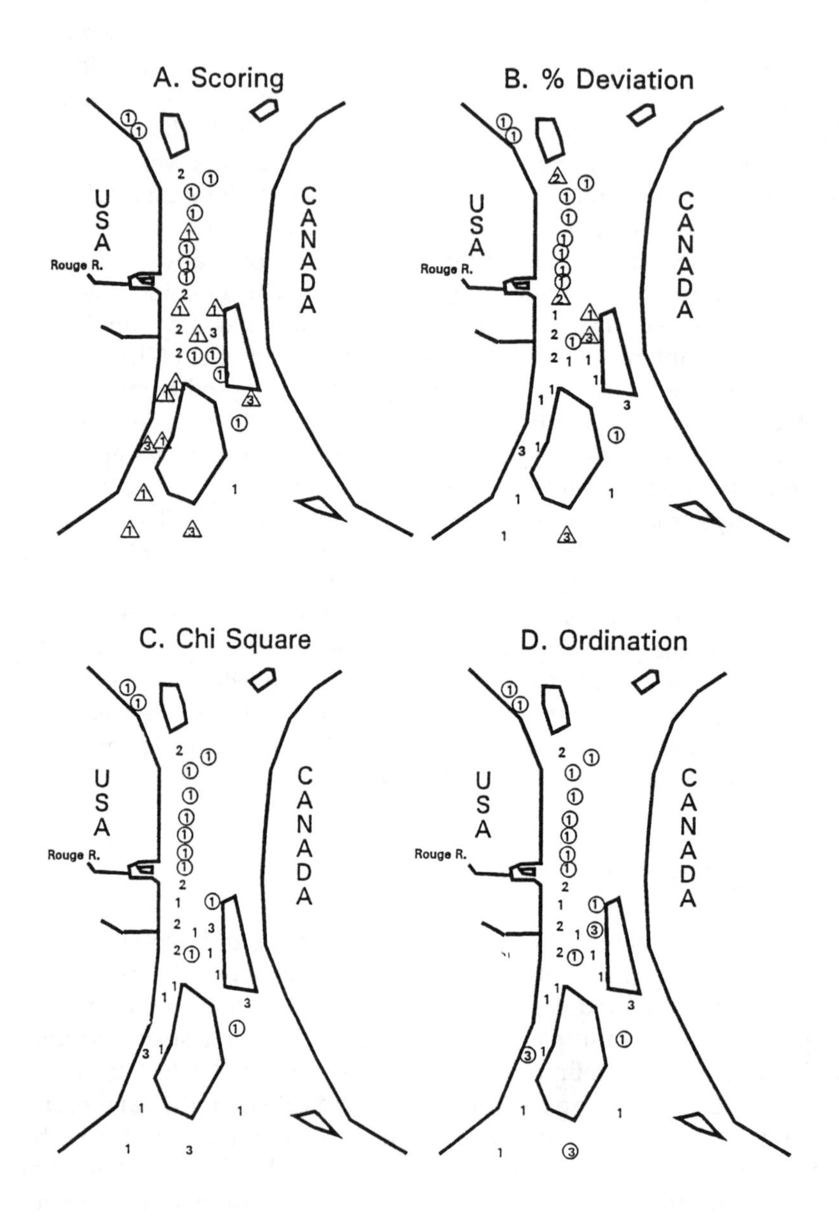

Figure 3: Predicted community (indicated by number) at 30 test sites in the Detroit River. Sites that met guidelines are circled, indecisive sites are indicated by a triangle and failed sites are unmarked.

Table 3: Performance of multiple discriminant analysis in predicting community assemblages at three levels of clustering and predicted cluster membership for reference sites for the three cluster solution.

Number of Clusters	Discriminant Statistics Probability of significance for the F statistic		Sites Correctly Predicted (%)
	Wilks Λ	Hotellings T^2	
3	.032	.026	83
4	.149	.115	59
8	.750	.740	35

Cluster	No. of Sites	Predicted Cluster Membership			Percent Correct Predictions
		1	2	3	
1	24	21	2	1	87.5
2	3	1	1	1	33.3
3	2	0	0	2	100

Guideline Development and Determination of Impairment

It is suggested that numerical biological guidelines should be based on the abundance and variance of those taxa that are most representative of the community assemblage (Table 2), or those taxa that are most abundant. For example, although the Isopoda, Hirudinoidea and Odonata are important in discriminating between the three communities (Table 2), they are relatively rare. On the other hand, the Amphipoda and Chironomidae, although less discriminating, are numerically more abundant and likely of more ecological importance. Four approaches to developing numerical guidelines for comparison of observed and expected communities are considered: (1) a simple scoring system; (2) percent difference in reference and test site means (± the standard deviation for each taxa from the reference community); (3) calculation of X^2 values and; (4) comparison in multivariate space. These approaches are examined below.

Scoring System

This is a simple and subjective method and is similar to the approach described by the USEPA (1990) in their rapid biological assessment and the approach used by Karr *et al.* (1986) in developing indices of biological integrity. The method assigns a score value to each taxon based upon its numerical importance and contribution to differentiating between communities. For example, the Ephemeroptera and Tubificidae were assigned a score of 25 as they were both abundant and discriminatory (Table 2). Other taxon were assigned a score of 10 as they were either rare or not as important in discriminating between assemblages. If the observed number of the taxon at a test site were present at numbers within 1 S.D. of the mean for the reference community (Table 2), then the full score for the taxon was assigned to that test site (either 25 or 10). If the numbers were higher than 1 S.D., then 5 was deducted from the score; if less than 1 S.D., then 0 was scored for the taxon. The scores are summed for each taxa and a site score assigned (maximum of 100). The scores can then be ranked, but there is no way of measuring the significance of differences and determination of impairment is subjective. For illustrative purposes, a score of 100-75 was considered a pass, 50- 75 as indecisive and 0-50 as a fail. The station scores are presented in Table 4. Of 22 sites predicted as being community 1, twelve met the guidelines, nine were indecisive and one failed. All the community 2 sites failed and the community 3 sites were indecisive. The sites that were deemed as unimpaired are largely those upstream of the Rouge River (Fig. 3); however, many of the downstream sites that have been described as having impacted communities (Reynoldson and Zarull 1989) are in the indecisive range. This result is an artifact of the scoring, which simply provides a ranking of the degree of difference. The method would require a great deal of data to calibrate and would always be open to criticism because of its subjective nature.

Mean Difference Method

This method is quantitative and provides a measure of absolute difference between a test site and the predicted community. Values are calculated as percent difference in the means for each taxon from the reference community assemblage, in two steps:

1. Calculate the percent difference in the means for each taxa.

(1)

$$\% Difference = \left(\left(\frac{x_{test}}{x_{reference}} \right) * 100 \right) - 100$$

2. Adjust for variation in reference community.

(2)

$$\% Difference = \% Difference_{(eqn\ 1)} - \% S.D._{(ref)}$$

Percent differences from Eqn. 2 are only counted if > 1, as scores of 0 or < 1 indicate the percent difference for that taxon are within 1 S.D. of the mean for the reference community and thus, should be counted as zero. The difference is calculated for each taxon and an average difference determined for the site.

The average percentage differences are shown in Table 4. Average differences of 10% or less are considered as meeting the guideline and differences of 100% or more can be categorized as failure. The intermediate range has to be considered as indecisive, depending on the management decision to be made. For example, if navigational dredging is being considered where open water disposal could have an impact, then

Table 4: Predicted community assemblages for 30 test sites in the Detroit river and performance of four proposed guidelines: score system, mean difference, chi-square and ordination (prinicipal component scores P.C.).

Site	Predicted Community	Score	Mean % Difference	χ^2	P.C. 1	P.C. 2	P.C. 3	Sum P.C.
D-1	1	100	0.0	0	0.34	0.13	0.17	0.65
D-15	1	100	0.0	0	0.40	0.13	0.34	0.87
D-5	1	100	0.0	0	0.33	0.10	0.28	0.71
D-E	1	100	0.0	0	0.20	0.04	0.12	0.36
D-F	1	100	0.0	0	0.52	0.18	0.37	1.07
D-10	1	75	0.1	0	0.51	0.18	0.35	1.04
D-17	1	75	0.1	0	0.33	0.10	0.26	0.69
D-37	1	75	0.1	0	0.54	0.18	0.39	1.10
D-19	1	75	1.5	0	0.49	0.16	0.34	0.99
D-53	1	75	1.5	0	0.53	0.18	0.37	1.08
D-13	1	70	4.1	0	0.08	0.00	0.15	0.23
D-27	1	70	58.8	0	0.45	0.00	0.31	0.77
D-46	1	70	206.1	47255	0.55	0.04	0.26	0.85
D-M	1	85	222.0	935	1.64	0.76	0.41	2.82
D-I	1	65	245.7	19469	0.56	0.10	0.31	0.96
D-36	1	65	609.8	43102	0.67	0.49	0.27	1.43
D-H	1	90	642.6	1424	1.30	1.29	0.49	3.08
D-23	1	65	735.1	62850	0.75	0.59	0.29	1.62
D-P	1	55	762.2	13237	0.06	1.20	2.55	3.81
D-51	1	45	870.0	19089	0.72	0.89	0.25	1.86
D-K	1	65	958.7	65258	0.74	0.89	0.21	1.84
D-50	1	70	2462.5	>100000	0.64	0.09	0.39	1.12
D-21	2	10	41.0	5710	1.20	0.48	1.60	3.28
D-9	2	10	84.0	23379	1.05	0.43	1.49	2.96
D-26	2	40	3831.9	>100000	1.96	2.10	1.86	5.93
D-34	2	35	5304.5	>100000	2.81	6.38	1.46	10.64
D-49	3	70	41.6	1369	1.89	0.82	2.81	5.51
D-35	3	60	84.9	1795	1.08	0.23	3.34	4.65
D-N	3	60	265.1	5902	0.69	1.00	3.65	5.34
D-O	3	60	295.0	>100000	2.08	0.71	1.59	4.38

erring on the side of safety is desirable and the indecisive (10-100%) sites should be considered as unsuitable. On the other hand, if remediation is being considered, which normally has high associated costs, then one would only consider those sites that clearly fail. With this system of guidelines, of the 22 community 1 sites, 11 pass, 10 fail and one is indecisive; for both communities 2 and 3, two sites fail and two sites are indecisive. There are fewer indecisive sites than with the scoring method and the distinction between sites up and down stream of the Rouge River is more evident (Figure 3).

Chi Squared Method

Using this predictive approach produces an expected community of key taxa (Table 2), based on the reference communities, and measured or observed numbers of the same taxa at the test site(s). Such data lend themselves to the determination of the probability of difference between reference and test sites using a chi-square test, thus allowing an estimation of the significance of the difference of the test site from its predicted state, with known probability. Chi-squared values were calculated for each station where the expected value for a taxon at a test site was the mean of the reference sites for that community, and the observed was the measured value for the same taxon at the test site. Chi-squared values were calculated for each of the key taxa (Table 2). The chi-squared was calculated using the mean value for the reference community (Table 2), yet the reference community has a variation associated with it, as expressed by the S.D. for each taxon. Therefore chi-squared was also calculated for the mean +1 S.D. and the calculated chi-squared for each test site adjusted by the chi-squared for the mean and 1 S.D. Adjusted values are presented in Table 4; 12 of the 22 community 1 sites were significantly different from the reference community. Those sites that supported the predicted community ($P > 0.05$) were upstream of the Rouge River, and all, but three, downstream stations were significantly ($P < 0.05$) impacted (Figure 3).

Multivariate Method

The final approach to comparing test sites to reference conditions used ordination methods. Both the reference sites and test sites were ordinated using PCA (Table 5). The results for test sites predicted as community 1 (Figure 4a) show a large number of test sites (solid squares) falling within the boundary of the reference sites (open squares) on the first two components, which explained over 60% of the variance. The degree of difference can be determined by the distance from the reference cluster centroid. For example, sites D-K, D-P and D-H are clearly different

Table 5: Concordance of four methods for developing sediment guidelines at 30 sites in the Detroit River (+ pass, - fail, 0 indecisive).

Site	Predicted Community	Score	Mean % Difference	χ^2	Ordination	Concordance %
D-1	1	+	+	+	+	100
D-15	1	+	+	+	+	100
D-5	1	+	+	+	+	100
D-E	1	+	+	+	+	100
D-F	1	+	+	+	+	100
D-10	1	+	+	+	+	100
D-17	1	+	+	+	+	100
D-37	1	+	+	+	+	100
D-19	1	+	+	+	+	100
D-53	1	+	+	+	+	100
D-13	1	+	+	+	+	100
D-27	1	0	0	+	+	50
D-46	1	0	-	-	-	75
D-M	1	+	-	-	-	75
D-I	1	0	-	-	-	75
D-36	1	0	-	-	-	75
D-H	1	+	-	-	-	75
D-23	1	0	-	-	-	75
D-P	1	0	-	-	-	75
D-51	1	-	-	-	-	100
D-K	1	0	-	-	-	75
D-50	1	0	-	-	-	75
D-21	2	-	0	-	-	75
D-9	2	-	0	-	-	75
D-26	2	-	-	-	-	100
D-34	2	-	-	-	-	100
D-49	3	0	0	-	+	0
D-35	3	0	0	-	+	0
D-N	3	0	-	-	-	75
D-O	3	0	-	-	+	50

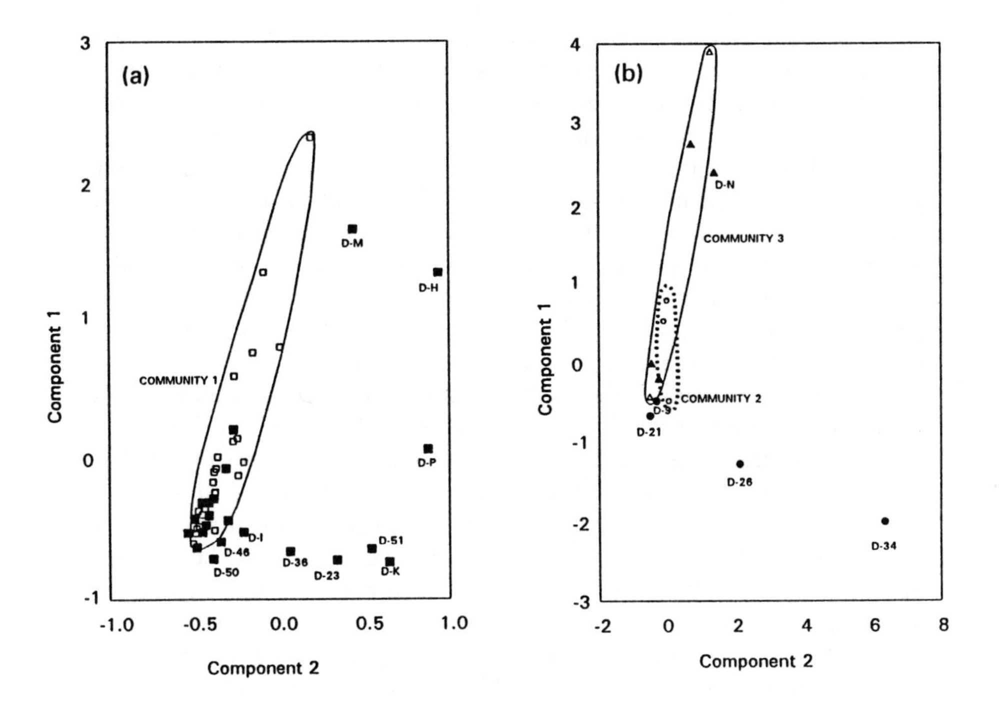

Figure 4: Ordination plots (PCA) for three reference communities (open symbols) and test sites (solid symbols). Reference community boundaries are hand-drawn. (a) Sites represented by community 1 (squares). (b) Sites represented by community 2 (circles) and 3 (triangles).

where site D-I is marginally different. Results are similarly shown for test sites predicted as being community 2 or 3 (Fig. 4b). Examination of the geographical location of the sites (Fig. 3) shows again that the upstream stations support the predicted communities and those in the vicinity and downstream of the Rouge River are different from that predicted.

Discussion

As far as we are aware, this is the first serious attempt to develop numerical, biological sediment guidelines. Other proposed biological objectives, guidelines or criteria have either been narrative statements (USEPA 1990) or more broadly ecosystemic (Hughes *et al.* 1986, Karr *et al.* 1986). While in the example of the Detroit River, we only include benthic invertebrate community structure data, this approach could easily be adapted to other biological data matrices. To develop sediment guidelines, reference sites should be classified based on both benthic invertebrate community structure and sediment bioassay response where the assay endpoints are the classifying variables, and a similar sediment data matrix is used to predict the response category. At a test site, the predicted response to the assay would be determined and measured against the actual response as described above.

The four proposed methods for actually developing guidelines show good concordance in this illustrative example (Table 5). Of the 30 test sites, 14 showed complete concordance. A further 10 sites showed 75% concordance between three of the four proposed systems. In eight of those 10 sites, the scoring system gave an indecisive result and in two sites the percent mean difference was indecisive. Five sites showed disagreement in whether the guidelines indicated pass or fail. In two cases (community 1), three of the guidelines indicated fail and the score system a pass, and in three cases (community 3), there was disagreement between ordination (pass) and two other methods. However, in the case of community 3, the reference community was only defined from two stations, and thus, the ordination boundaries of that community are imprecise. There is general agreement between all the methods, which is particularly surprising given the subjective nature of the scoring system and the subjective decision points for the mean difference approach. The two more favoured methods are the chi-squared, which is quantitative and objective, and the ordination technique which integrates all the available information from the sites and places the test sites in the same environment as the reference sites. The ordination technique also provides a clear pass or fail criterion, either within or outside the reference community boundaries, and a measure of the extent of failure by the degree to which the test site is outside the reference community boundary.

The Detroit River example is provided for illustrative purposes only. Application of such an approach would require a much larger data set of reference sites. A river classification project in the U.K. (Wright *et al.* 1984) initially used 268 sites to produce 16 community assemblages; subsequently, the database was expanded to included 370 sites defining 30 community assemblages (Armitage *et al.* 1987). In Canada, a project currently in progress (Reynoldson and Day 1991) is assembling a data set of 250 reference sites for the Great Lakes using both community structure and sediment toxicity. This initially large effort is only required once and can be gradually augmented to and refined as data are acquired. This fundamental approach of reference site classification, development of a predictive model, prediction of test sites and comparison with guidelines derived from the reference sites can be used in any type of environment for estimation of ecological integrity. The approach simply requires the selection of appropriate biological data matrices. It will not replace a chemically based approach, particularly for "end of pipe" application, but does resolve many of the inherent problems associated with chemical guidelines, objectives and criteria.

ACKNOWLEDGEMENTS

We would like to thank S. Thornley and Y. Hamdy for providing the test data for the Detroit River, as well as, Jim Kay and particularly, Peter Chapman for their review comments.

REFERENCES

Adams, D.D. and D.A. Darby. 1980. A dilution mixing model for dredged sediments in freshwater ecosystems. In: R.A. Baker (ed.) *Contaminants and Sediments*. Vol. 1. Ann Arbor Sc. Publishers Inc. Ann Arbor, Mi. pp. 373-392.

Armitage, P.D., R.J.M. Gunn, M.T. Furse, J.F. Wright andD. Moss. 1987. The use of prediction to assess macroinvertebrate response to river regulation. *Hydrobiol.* 144:25-32.

Beak Consultants Ltd. 1987. Development of sediment quality objectives Phase 1 Options. Report prepared for the Ontario Min. Environment.

Cairns, J. Jr., D.L. Kuhn and J. L. Plafkin. 1979. Protozoan colonization of artificial substrates. In: R.L. Wetzel, ed. *Methods and Measurements of Materials.* Philadelphia, PA, pp. 39-51.

Chapman, P.M. 1986. Sediment quality criteria from the sediment quality triad: an example. *Environ. Toxicol. Chem.* 5:957-964.

Chapman, P.M. and E.R. Long. 1983. The use of bioassays as part of a comprehensive approach to marine pollution assessment. *Marine Pollution Bulleti.n* 14: 81-84.

Corkum, L.D. and D.C. Currie. 1987. Distributional patterns of immature Simuliidae (Diptera) in northwestern. *N. America. Freshw. Biol.* 17:201-221.

Dillon, T.M. and A.B. Gibson. 1986. Bioassessment methodologies for the regulatory testing of freshwater dredged material. Proceedings of a workshop. Miscellaneous Paper EL-86-6. Department of the Army, US Army Engineer Waterways Experiment Station, Vicksburg, MS.

Giesy, J.P. and R.A. Hoke. 1990. Freshwater sediment quality criteria: Toxicity bioassessment In: *Sediments: Chemistry and Toxicity of In-Place Pollutants.* pp. 265-348. Eds. R. Baudo, J. Giesy and H. Muntau. Lewis Publishers Inc., Ann Arbor, MI.

van de Guchte, C. and C.J. van Leeuwen. 1988. Sediment Pollution. In: *Manual on Aquatic Ecotoxicology* (eds.) de Kruijf, H.A.M., de Zwart, D., Ray, P.K. and Viswanathan, P.N. Kluwer Acad. Dordrecht, Neth.

Hughes, R.M., D.P. Larsen and J.M. Omernik. 1986. Regional reference sites: a method for assessing stream pollution. *Environ. Manage.* 9:253-262.

International Joint Commission (IJC). 1987. *Guidance on Characterization of Toxic Substances Problems in Areas of Concern in the Great Lakes Basin.* A Report from the Surveillance Work Group, Windsor, Ontario, March 1987. 177pp.

International Joint Commission (IJC). 1988a. *Procedures for the Assessment of Contaminated Sediment Problems in the Great Lakes.* A Report by the Sediment Subcommittee, Windsor, Ontario 140pp.

International Joint Commission (IJC). 1988b. *Procedures for the Remediation of Contaminated Sediment Problems in the Great Lakes.* A Report by the Sediment Subcommittee, Windsor, Ontario 140pp.

Johnson, R.K. and T. Wiederhol. 1989. Classification and ordination of profundal macroinvertebrate communities in nutrient poor, oligo-mesohumic lakes in relation to environmental data. *Freshw. Biol.* 21:375-386.

Karr, J.R., P.L. Fuasch, P.L. Angermier, P.R. Yant and I.J. Schlosser. 1986. Assessing biological integrity in running waters: a method and its rationale. Spec. Publ. 5. Ill. *Nat. Hist. Surv.*

King, J.M. 1981. The distribution of invertebrate communities in a small South African river. *Hydrobiol.* 83:43-65

Long, E.R. and P.M. Chapman. 1985. A Sediment Quality Triad: measures of sediment contamination, toxicity and infaunal community composition in Puget Sound. *Marine Pollution Bulletin.* 16(10): 405-415.

Mathews, R.A., A.L. Jr. Buikema, J. Jr. Cairns, and J.H. Jr. Rodgers. 1982. Biological monitoring Part IIA - Receiving system functional methods, relationships and indicies. *Water Res.* 16:129-139.

Metcalfe, J.L. 1989. Biological water quality assessment of running waters based on macroinvertebrate communities: History and present status in Europe. *Environ. Poll.* 60:101-139.

Moss, D., M.T. Furse, J.F. Wright, and P.D. Armitage. 1987. The prediction of the macro-invertebrate fauna of unpolluted running water sites in Great Britain using environmental data. *Freshw. Biol.* 17:41-52.

Ministry of Transport and Public Works. 1989. Water in the Netherlands: A time for action. National Policy Document on Water Management. Min. Trans. Pub. Wks., Hague, Neth. 68pp.

Ontario Min. Environment. 1988. Guidelines for the management of dredged material. Ont. Min. Envir., Toronto, Canada.

Ormerod, S.J. and R.W. Edwards. 1987. The ordination and classification of macroinvertebrate assemblages in the catchment of the River Wye in relation to environmental factors. *Freshw. Biol.* 17:533-546.

Persaud, D., R. Jaagumagi and A. Hayton. 1989. Development of provincial sediment quality guidelines. Ont. Min. Env., Toronto, Canada. 19pp.

Reynoldson, T.B. and K.E. Day. 1991. A study plan for the devolopment of biological sediment quality guidelines. NWRI unpub. rept. Burlington, Ont. 35 pp.

Reynoldson, T.B. and M.A. Zarull. 1989. The biological assessment of contaminated sediments - the Detroit River example. *Hydrobiol.* 188/189:463-476.

Thornley, S. and Y. Hamdy. 1984. An assessment of the bottom fauna and sediments of the Detroit River. Ontario Min. Environment. 48 pp.

United States Environmental Protection Agency (USEPA). 1977. Guidelines for the classification of Great Lakes harbor sediments. *Region V.* Chicago, Illinois, 7pp.

United States Environmental Protection Agency (USEPA). 1989. Sediment classification methods compendium. Unpublished Report.

United States Environmental Protection Agency (USEPA). 1990. *Biological Criteria.* EPA-440/5-90-004 Washington D.C. 57 pp.

Wilkinson, L. 1990. SYSTAT: The system for statistics. SYSTAT Inc. Evanston, Ill.

Wright, J.F., D. Moss, P.D. Armitage, and M.T. Furse. 1984. A preliminary classification of running water sites in Great Britain based on macro-invertebrate species and the prediction of community type using environmental data. *Freshw. Biol.* 14:221-256.

On the Nature of Ecological Integrity: Some Closing Comments

JAMES J. KAY
Environment and Resource Studies, University of Waterloo
Waterloo, Ontario

If we are to be stewards of the ecological systems that make up our biosphere, then we must have some state, in which we wish these systems to be. The term 'integrity' has become the name we use for this state. Environmental managers find themselves mandated by various policy statements and legislation (the Great Lakes Water Quality Agreement 1978, the Canada Park Service Act 1988, the draft Montana Environmental Protection Act 1992, Environment Canada's mission statement 1992) to preserve/maintain/promote/protect/restore ecological integrity. The challenge now is to define "integrity."

Regier, in his chapter, explores the notion of integrity. For some, at first glance, his exploration will seem too abstract and imprecise to be helpful in the "real world." In essence, he says that a system has ecological integrity when it is perceived to be a whole that is in a state of "well being." He explores what this means for different groups of people, in different situations, with different perspectives. This definition is as specific as a general definition of ecological integrity can be. Our sense of what constitutes ecological integrity is very much dependant on our perspective of what constitutes a whole ecological system. When we are dealing with ecological systems, we are dealing with complex systems, systems which require complex descriptions that are case specific. Simple answers, definitions and hypotheses (like species diversity leads to ecological stability) are never adequate for discussing complex systems and, by implication, ecological integrity.

This is not to say that there are no generic statements to be made. King, in his chapter, acquaints us with many of the lessons to be learned from complex systems theory with regard to the issues that must always be of concern in any inquiry into ecological integrity. If ecological integrity refers to our sense of the wholeness and well being of an ecological system, then our assessment of integrity should begin with an analysis of the system we are examining.

Systems analysis always begins by considering spatial and temporal scale and hierarchy. Are we concerned with the integrity of a landscape, a forest, a particular species or all of these? Over what time scale and what spatial extent? What are the important processes which make up the system? To answer these questions, as King discusses, is to define the perspective we are taking on, to separate the things which we deem important from those we think unimportant.

One of the lessons of systems theory is that there is no preferred observer. Thus, we must be careful to explicitly specify the system, (i.e., identify scale, hierarchy, boundaries, system environment, etc.) because part of this process is the identification of the issues of importance, that is, the contextual perspective for the integrity evaluation. Neglecting this step will implicitly impose a perspective, including a set of values, on the integrity evaluation. A consequence of this is that the contextual issues will not be thought through, but rather implicitly assumed. The issues of importance and relevance will not be explicitly identified and the experience of systems theory tells us that they will likely be neglected.

Furthermore, specifying the system means identifying not only the physical and biological issues that scientists feel are important — and given the state of the science of ecology, there is much debate about these (Kay, 1991b) — but also identifying the important socio-economic, political and policy issues. Several authors in this volume (Regier, King, Steedman and Haider, Munn, and Marshall *et al.*) have noted that it is crucial to explicitly include societal issues and values in any evaluation of ecological integrity. Otherwise, the evaluation will only reflect the values and concerns of the evaluator, and as such, is likely to have little meaning, or be of little use, to policy and decision makers (who must represent a broad group of stakeholders).

Many environmental managers are reluctant to include societal issues and values in an evaluation of ecological integrity. As Steedman and Haider point out, their training is usually in the physical and biological sciences, a training which instills a reluctance to sully any analysis with fuzzy, subjective, qualitative issues, as this would only serve to dilute the "objectivity" of science with the subjectivity of values. However, reality is that when one specifies systems, one is not dealing with objective science, but with perspectives, with ways of looking at the world, and these always reflect a value system. We must acknowledge this reality and make it front and center in our evaluations of integrity. This theme echoes throughout this book— that a discussion of ecological integrity without a discussion of the social, economic, political and policy concerns is not a meaningful discussion.

So where does this leave us? It is not surprising, in light of this discussion, that attempts to develop general criteria for evaluating, and thereby defining, ecological integrity have not been successful. Clearly, integrity is context specific, the context being the physical, biological, social and cultural features of the specific geographic area in question. Furthermore, an evaluation of integrity must look at issues from different hierarchical and scale perspectives. Thus, integrity can only be defined clearly (in terms of evaluative criteria) for specific ecosystems, in the context of the humans which are an integral part of the ecosystem. In no way does this detract from the concept of integrity. This is just the reality of dealing with complex systems.

Furthermore, evaluating and defining ecological integrity can never be seen as purely an exercise in the physical and biological sciences. Ecological integrity is about our sense of the wholeness and well being of ecological systems and, in this, must reflect our sense of what we value in them.

So what are the roles of the physical and biological sciences in discussing ecological integrity? Serafin and Steedman address this issues as follows:

> Meaningful discussion of ecological integrity must be tempered by the realization that the concept does not exist outside of human value judgements, unlike notions such as gravity or general relativity. Measurements of integrity by scientific means are therefore limited to criteria which are subjectively selected, even though those criteria are frequently paraded as objective and value-free. Can the methods of measuring integrity be scientific even though the underlying premise is not?
>
> The nature of the dilemma is glaring if we consider a question like: Did the Mount St. Helen or the Krakatoa events cause an impairment of ecological integrity? If we say no, implying that nature can do no wrong, we have made a definite value judgement. Conversely, arguing that the biotic aspect of the ecosystem was degraded invites a rebuttal built around forest fires, floods, or other dramatic stochastic events. And again we have made a value judgement. Therefore, it seems apparent that we can measure and analyze CHANGES in an ecosystem, but we can only make JUDGEMENTS about the integrity of that system.

Complex systems theory has much to say about the kind of changes to expect in a developing ecological system. Such systems are described as nonlinear, which is to say they behave as a connected whole. The system's behaviour cannot be explained by deconstructing it into pieces, pieces whose individual behaviour can be linearly summed together to give the behaviour of the whole (system). Thus, one cannot understand the behaviour of ecosystems from examining only the behaviour of the individuals and species within it. Rather a hierarchical approach is required (Allan and Starr 1982, Allan and Hockstra 1992).

Complex systems have multiple steady states. Thus, in a specific geographical location, there is the potential for a number of different ecosystems, communities and species to exist, in addition to the ones currently present. It is possible that such systems may only be dynamically stable, that is a stable equilibrium point does not exist for the system. Holling (1986, 1992) gives a number of examples of this and suggests that the notion of a steady state is really an illusion. Ecosystems are continually undergoing a birth-growth-maturity-death process which he refers to as an exploitation, conservation, release, reorganization cycle. Interrupting this cycle through, for example forest fire suppression, not only interferes with the normal cycle of life but also magnifies the magnitude of changes when they are eventually triggered.

Movement through the cyclic process, described by Holling, is not continuous. There are temporary stationary states. The process is often characterized by bifurcations and flips, that is, catastrophic behaviour. Furthermore, elements of the process can be chaotic, that is, inherently unpredictable. I have explored the implications of these behaviours for ecosystem integrity elsewhere (Kay 1991a). The main point of this work is that the normal behaviour of ecological systems is quite complex and the simple notion of succession to a stable climax community is not sufficient. Nor is the notion that loss of ecological integrity corresponds to disturbance of the ecosystem away from a climax. Life is not about maintaining stable equilibrium states. It is dynamic and ever changing.

The challenge facing us is to discover what rules, if any, govern the overall direction of ecosystem development and ecosystem reorganization induced by environmental change. Ecosystems exhibit chaotic and catastrophic behaviour and this, combined with the inherent variability in their organizational processes, means that we will always have uncertainty about their dynamic behaviour. Ecological systems belong to a class of systems Weinberg refers to as middle number systems. The main issue is to discover what constraints, if any, there are on the behaviour of these systems. Is there an overall direction to development and reorganization in response to environmental change? How can development be characterized? Can we decrease our uncertainty about developmental and reorganization processes and by how much?

Currently, there are a number of investigators working on these issues. (See Gunther and Folke, 1992 for a review.) We (Schneider and Kay 1993) have proposed a nonequilibrium thermodynamic basis for understanding, as self-organizing phenomena, ecosystem development and response to environmental change. The development of self-organizing systems is characterized by periods of rapid organization followed by an interval during which the system maintains itself in a steady state. Organization of the system is not a smooth process but rather proceeds in spurts. Each spurt results in the system moving further from equilibrium, dissipating more energy, and becoming more organized. Ecosystem succession is an example of this kind of process. Each of the seral stages corresponds to one steady-state plateau. The displacement of a previous seral stage by the next is an example of a spurt, the re-organization of the system to a new level of structure which dissipates more energy.

As ecosystems develop, they become more organized and effective at dissipating solar energy (Kay, 1984). But these self-organizing tendencies are limited by external environmental fluctuations that tend to disorganize ecosystems. The point in state space where the disorganizing forces of external environmental change and the organizing thermodynamic forces are balanced is referred to as the "optimum operating point." The climax community in ecological succession would be an example of an ecosystem's optimum operating point. The climax community represents a temporary balance between the organizing and the disorganizing forces in ecosystems.

Our sense of the system as a whole, that is its integrity, has to do with its ability to maintain its organization and to continue its process of self-organization. If a system is unable to maintain its organization, in the face of changing environmental conditions, then it has lost its integrity. In essence, integrity has to do with the ability of the system to attain and maintain its optimum operating point.

Integrity, that is our sense of an ecosystem as a whole, encompasses three major ecosystem organizational facets. Ecosystem health, the ability to maintain an optimum operating point under normal environmental conditions, is the first requisite for ecosystem integrity. But it alone is not sufficient. An ecosystem must also be able to cope with changes in environmental conditions, that is, stress. Thus in order to have integrity, an ecosystem must be able to attain and maintain an optimum operating point when stressed. (This optimum operating point may be different from the unstressed optimum operating point). Finally, an ecosystem which has integrity must be able to continue evolving and developing, that is, continue the process of self-organization on an ongoing basis. It is these latter two facets of ecosystem integrity that differentiate it from the notion of ecosystem health.

Our understanding of the behaviour of complex self-organizing systems provides a framework for the investigation of environmentally induced changes in ecosystem organization and integrity (Kay 1991a). Table 1 summarizes the potential changes that can occur in an ecosystem's organization, if environmental conditions change. (The environmental change may be short term with the environment returning to its previous condition, or the change may persist.) This analysis establishes that ecosystems can respond to changes in the environment in qualitatively different ways. One response is for the system to continue to operate as before, even though its operations may be initially and temporarily unsettled. A second response is for the system to operate at a different level using the same structures it originally had (for example, a reduction or increase in species numbers). A third response is for some new structures to emerge in the system that replace or augment existing structures (for example, new species or paths in the food web). A fourth response is for a new ecosystem, made up of quite different structures, to emerge. This enumeration, of possible ecosystem responses to environmental change, is far richer than the simple classical notion that stress temporarily displaces an ecosystem from its climax, to which it hopefully returns. It also points out that an ecosystem has no single preferred state for which it should be managed.

This discussion identifies ways in which an ecosystem might reorganize in the face of environmental change, but not which re-organization constitutes a loss of integrity. It could be argued (and often is) that any environmental change, which permanently changes the optimum operating point, affects the integrity of the ecosystem. If this is correct, then an ecosystem would have integrity only if it has the ability to absorb environmental change without any permanent ecosystem change. Thus the other four distinct ecosystem responses to environmental change (Cases 0 through 3 in Table 1) would constitute a loss of integrity, even though the latter three are responses in which the ecosystem reorganizes itself to mitigate the environmental change. This is not reasonable since the reorganized ecosystem is usually just as healthy as the original, even though it may be different. There is no *prima facie* scientific reason that an existing ecosystem should be considered to be the only one to have integrity in a situation, just because of its primacy.

It could also be argued, that any system that can maintain itself at any optimum operating point, has integrity. In this case, loss of integrity

Table 1: The possible responses of an ecosystem to environmental change (after Table 1, Kay, 1991a).

a) The ecosystem does not move from its original optimum operating point.

b) The ecosystem moves from its original optimum operating point but returns to it.

Issues concerning integrity to be considered for this case:

1) How far is the system moved from its optimum operating point before returning?
2) How long will it take to return to its optimum operating point?
3) What is the stability of the system upon its return?

c) The ecosystem moves permanently from its original optimum operating point.

Case 0: The ecosystem collapses.
Case 1: The ecosystem remains on the original developmental pathway.
Case 2: The ecosystem bifurcates from the original developmental pathway to a new one.
Case 3: The ecosystem moves catastrophically to a different developmental pathway.

Issues concerning integrity to be considered for each of these four cases:

1) How far is the new optimum operating point from the old?
2) How long does it take to reach the new optimum operating point?
3) What is the stability of the system about the new optimum operating point?
4) If the environmental conditions return to their original state, will the system return to the original optimum operating point?

would occur only if the system is unable to maintain itself at an optimum operating point (Case 0 in Table 1). Fortunately this is rarely the case, desertification being one of the few examples.

Neither of these definitions of integrity is operationally useful. The former definition is too restrictive, in most situations, and reflects a desire to preserve the world exactly as it is. This denies the fundamental dynamic nature of ecosystems. It leads to disastrous mismanagement, e.g., the complete suppression of forest fires which eventually results in catastrophic conflagrations. (Of course one may wish to preserve an ecosystem as an example or specimen of a specific type.) The latter definition does not help managers as it restricts loss of integrity to a situation which rarely occurs and which is clearly undesirable. Hence, this definition would be trivial.

A definition, that is between these two extremes, is to specify that some changes in optimum operating points are undesirable, and therefore represent a loss of integrity. But this would require a value judgement about which changes are desirable and which are undesirable. The science of ecology can, in principle, inform us about the kind of ecosystem response or reorganization to expect in a given situation. But it does not provide us with a scientific basis for deciding that one change is better than another, except possibly in the two extreme cases just discussed. So again the insight into ecological integrity, gained from complex systems theory, is that the physical and biological sciences can describe, and even predict, changes in the biosphere, but they alone cannot determine which is better. Ultimately an evaluation of the ecological acceptability of a human activity will depend on a value judgement about whether the resulting changes in the effected ecosystem are acceptable to the human participants.

In the final analysis, to define ecological integrity is to define a set of ecological characteristics to be monitored for change beyond specific values. To operationalize the notion of integrity requires the development of a monitoring framework and its associated measures and indicators. Much of this book is devoted to this question and only some of the highlights will be repeated here.

There are three distinct issues to be dealt with in developing a monitoring scheme for ecological integrity. First, the changes in organization to be associated with changes in ecological integrity must be identified. This issue is naturally raised as part of the system's specification exercise discussed earlier. Then comes the scientific and technical problem of how to quantify these changes. What needs to be measured and how is it best done? These are scientific and technical issues with which we are all struggling. Once measurements have been decided upon, the issue of evaluation becomes paramount. For what values of the measures will integrity be deemed to have been lost? Who will make this decision and who will act on it? The first and the last of these three issues require the understanding of physical and biological scientists to be combined with the concerns of society as voiced by elected officials, policy makers, public interest groups, and others. Only when this combination occurs, will it be reasonable to expect ecological integrity to result from our stewardship of the biosphere.

The process of developing a monitoring framework must begin with identifying the users of the information gained from the monitoring. What information do they need, and in what form must it be presented to be viable? With this established, a system identification exercise is conducted so as to resolve such issues as what hierarchical levels will be focused on, what temporal and spatial scales will be covered, and what processes need to be monitored.

Measures for several different hierarchical levels and scales will be needed (Shackell and Freedman, King). The measures will be rooted in the bioregion and social issues in question (Keddy *et al.*). Also, measures drawn from a number of different theoretical perspectives (e.g., landscape ecology, self-organization theory, population biology, etc.) will be required for a complete picture (Karr). Some measures will monitor the general condition of the ecosystem and its environment. Others will focus on specific known threats and the system's response to these threats. (Marshall *et al.* Woodley). Some will assess damage while others will serve as early warning alarms.

Clearly, a variety of measures are necessary to adequately monitor ecological integrity. King points out that measures must not be over-integrated as this will result in the loss of valuable information. Steedman and Haider, and Keddy *et al.* suggest methods for developing monitoring schemes and indicators. Woodley, Shackell and Freedman, Marshall *et al.* and Munn all discuss criteria for monitoring programs and indicators of ecological integrity. The work of Scheifer *et al.* (1988) and Harris *et al.* (1987) also provide guidelines for selecting measures. Recently, the IJC has published a framework for developing indicators of ecosystem health for the Great Lakes Region (International Joint Commission 1991).

Having completed the process of identifying the changes to be monitored and the measure thereof, we still face formidable technical and scientific problems. Ecology is still a young science and many methodological issues remain to be resolved. Karr points out, for example, that there is a tendency to take averages as the signal and variability in what is measured, as noise. Both he and Woodley point out that the variability is often, in fact, the signal. But how to measure this?

The next step is development of evaluation criteria. Steedman and Haider propose a process for developing criteria in general, while other authors examine this issue in more specific contexts. A main problem in this respect is the lack of baseline data for comparison purposes. In this regard, Rubec and Marshall identify three priorities for Canada:

1. A first order of business must be *integrating and synthesizing human activity and ecoregional information across Canada.*
2. Development of national measures of ecosystem health and integrity will also require the *creation of a national network of stable, secured and intercomparable benchmark reference sites with at least one such site in all ecoregions.*
3. We must ensure that *existing interdisciplinary monitoring sites and networks are maintained* (such as the Experimental Lakes Area, and the Dorset, Turkey Lakes, Lac Laflamme and Kejimkujik Watersheds developed in the national acid rain program) to provide the long term perspective essential to monitoring and uniquely available through these areas.

Examples of monitoring schemes and specific indicators of integrity are presented in this book: for bottom sediments (Reynoldson and Zarrell), for wetlands (Keddy *et al.*), for the Atlantic ecozone (Schackell and Freeman), for stream ecosystems (Karr), and for national parks (Woodley). Kay and Schneider (1993) discusses measures of ecosystem integrity based on food web analysis. All of these discussions of ecological integrity are rooted in a specific context.

Complex systems theory and traditional science can help us describe and understand changes in ecological systems. Systems theory can help us focus on issues of importance, vis à vis integrity. However, they alone cannot determine which ecological changes constitute a loss of integrity. When we define ecological integrity, we are undertaking to integrate everything we know about an ecological system and where we want it to be. This integration, to be complete, must include the sum total of human preferences and concerns about the system. We must find paths for ecosystem development which assure our species survival both in the short and long term. Such paths will balance the needs of other species with our own, so as to maintain a biosphere in which humans have a sustainable niche.

ACKNOWLEDGEMENTS

Prof. Henry Regier and Prof. George Francis have discussed these ideas at length with me and included me in a number of workshops and meetings which they have held on the topic. Dr. Raf Serafin provided us with a written commentary on the workshop and discussions which I have drawn upon herein. Discussions with Prof. L. Westra helped focused my thinking on the non-scientific issues surrounding ecological integrity.

REFERENCES

Allen, T. F. H., T.W. Hoekstra. 1992. *Toward a Unified Ecology*. New York: Columbia University Press.

Allen, T. F. H., T.B. Starr. 1982. *Hierarchy: Perspectives for Ecological Complexity*. University of Chicago Press.

Günther, F., C. Folke. 1992. *Nested Living Systems*. Stockholm: Royal Swedish Academy of Science. 22 pp. (Beijer Discussion Paper Series #5).

Harris, H. J. [and others]. 1987. Coupling Ecosystem Science with Management: A Great Lakes Perspective from Green Bay. Lake Michigan, USA. *Env. Mgmt.* 11(5): 619-625.

Holling, C.S. 1986. The Resilience of Terrestrial Ecosystems: Local Surprise and Global Change. In: Clark, W.M., Munn, R.E. (eds.). *Sustainable Development in the Biosphere*. Oxford University. pp. 292-320.

Holling, C.S. 1992. Cross-scale Morphology, Geometry, and Dynamics of Ecosystems, *Ecological Monographs*, in press.

International Joint Commission. 1991. *A Proposed Framework for Developing Indicators of Ecosystem Health for the Great Lakes Region*. 47 pp.

Kay, James. 1984. Self-Organization in Living Systems. Ph.D. thesis, Systems Design Engineering, University of Waterloo; Waterloo, Ontario, Canada.

Kay, J.J. 1991a. A Non-equilibrium Thermodynamic Framework for Discussing Ecosystem Integrity, *Environmental Management*. Vol 15, No.4, pp.483-495.

Kay, J.J. 1991b. The Concept of "Ecological Integrity," Alternative Theories of Ecology, and Implications for Decision-Support Indicators. In: Canadian Environmental Advisory Council *Economic, Ecological, and Decision Theories: Indicators of Ecologically Sustainable Development*. pp. 22-58 Commissioned background paper prepared for Canadian Environmental Advisory Council workshop on *Indicators for Ecologically Sustainable Development*. 25-26 July 1990, Ottawa, Canada.

Kay, J.J., E.D. Schneider. 1993. Thermodynamics and Measures of Ecosystem Integrity. In: *Proceedings of the International Symposium on Ecological Indicators*. Fort Lauderdale, Florida, Elsevier.(in press) 13 pp.

Schaeffer, David, J. E. Herricks, H. Kerster. 1988. Ecosystem Health: I. Measuring Ecosystem Health, *Env. Mgmt.* 12(4): 445-455.

Schneider, E.D, J.J. Kay. 1993. "Life as a Manifestation of the Second Law of Thermodynamics" (in press), *Advances in Mathematics and Computers in Medicine*. 30 pp.

Serafin, R., R. Steedman. 1991. "Working Group Report: Measuring Integrity at the Municipal Level," from a workshop on *Ecological Integrity and the Management of Ecosystems*. Heritage Resource Center, University of Waterloo; Waterloo, Ontario, Canada.

Weinberg, Gerald M. 1975. *An Introduction to General Systems Thinking*. John Wiley and Sons.

Appendix

List of Contributors & Advisors

List of Contributors & Advisors

Don Alexander
School of Urban and Regional Planning
University of Waterloo

John D. Ambrose
The Arboretum
University of Guelph

John Argalis
Environment and Resource Studies
University of Waterloo

James Baker
Wildlife Policy Branch
Ministry of Natural Resources
Toronto, Ontario

Bruce L. Bandurski
International Joint Commission
Washington, D.C. U.S.A.

Lawrence Baschak
School of Landscape Architecture
University of Guelph

Karen Beazley
The Landplan Collaborative Ltd.
Guelph, Ontario

R.W. Bethke
Department of Zoology
University of Guelph

A.C. Bird
Canadian Parks Service
Charlottetown, P.E.I.

Stuart Carolyn
Niagara Escarpment Commission
Georgetown, Ontario

John Carruthers
Policy Advisor
Program Management Directorate
Canadian Parks Service
Ottawa, Ontario

Susan Crispin
Great Lakes Heritage Data Network
c/o Nature Conservancy of Canada
Toronto, Ontario

Dan Daly
City of Kitchener
Parks Operations
Kitchener, Ontario

Peter Deering
Canadian Parks Service
Point Pelee National Park
Leamington, Ontario

Dr. David DeYoe
Ontario Ministry of Natural Resources
Forest Management Branch
Sault Ste. Marie, Ontario

Dr. George Dixon
Chairman
Biology Department
University of Waterloo

Cliff Drysdale
Canadian Parks Service
Kejimkujik National Park
Maitland Bridge, Nova Scotia

Dennis Dubé
Forestry Canada
Hull, Quebec

Alain Dufresne
Canadian Parks Service
Haute-Ville, Quebec

Dr. Peter Eaton
Environment Canada
Dartmouth, Nova Scotia

Harold Eidsvik
Senior Policy Advisor
Program Management Directorate
Canadian Parks Service
Ottawa, Ontario

David Euler
Ministry of Natural Resources
Fenwick, Ontario

Greg Filyk
Wildlife Habitat Canada
Ottawa, Ontario

Max Finkelstein
Canadian Parks Service
Hull, P.Q.

John Fisher
Trent University
Peterborough, Ontario

Luc Foisy
Canadian Parks Service
Quebec Region
Quebec

Graham Forbes
Department of Geography
University of Waterloo

Robert France
McGill University
Kingston, Ontario

George Francis
Environment and Resource Studies
University of Waterloo

Dr. Bill Freedman
School of Resource and
Environmental Studies
Dalhousie University
Halifax, Nova Scotia

Ray Frey
Environment Canada, Parks
Service
Riding Mountain National Park
Wasagaming, Manitoba

Lyle Friesen
University of Waterloo

Michael Gilbertson
International Joint Commission
Windsor, Ontario

Chris Gosselin
Regional Municipality of Waterloo
Dept. of Planning & Development
Waterloo, Ontario

Nel Grond
Landscape Architecture
University of Guelph

Bryon Hall
Ministry of Natural Resources
Temagami, Ontario

Michael Hall
Ontario Ministry of Natural Resources
Sudbury, Ontario

Dr. Peter Hall
Science Directorate
Forestry Canada
Ottawa, Ontario

Veena Halliwell
Canadian Environmental Advisory Council (CEAC)
Ottawa, Ontario

Sherry Hambly
Ontario Ministry of Natural Resources
Northeastern Region
Sudbury, Ontario

Douglas Harvey
Senior Planner
Northern Parks Establishment
Canadian Parks Service
Ottawa, Ontario

David Heaman
Ministry of Natural Resources
Temagami, Ontario

Brian Hindley
Ministry of Natural Resources

Rob Home
City of Cambridge
Cambridge, Ontario

Colleen Hyslop
Canadian Wildlife Service
Ottawa, Ontario

Scott Jones
Ministry of Natural Resources
Sault Ste. Marie, Ontario

James Karr
Department of Biology
Virginia Polytechnical Institute
Blacksburg, Va. U.S.A.

James Kay
Environment and Resource Studies
University of Waterloo

Dr. Paul Keddy
Department of Biology
University of Ottawa

Will Kershaw
Ministry of Natural Resources
Sudbury, Ontario

Dr. Tony King
Environmental Science Division
Oak Ridge National Laboratory
Oak Ridge, TN U.S.A.

Maria Kothbauer
Geography Dept.
Wilfrid Laurier University
Waterloo, Ontario

Daniel Kraus
Dept. of Environment
Canadian Parks Service
Cornwall, Ontario

Harold Lee
University of Ottawa

Nik Lopoukhine
Ecologist
National Parks Directorate
Canadian Parks Service
Hull, Quebec

Samm MacKay
K-W Field Naturalists
Waterloo, Ontario

Ian Marshall
State of Environment Reporting Branch
Environment Canada
Ottawa, Ontario

Virgil Martin
Department of Independent Studies
University of Waterloo

Ron Maslin
Canadian Parks Service

George Mercer
Canadian Parks Service
Wood Buffalo National Park
Fort Smith, N.W.T.

Greg Michalenko
Environment and Resource Studies
University of Waterloo

Robert J. Milne
Department of Geography
University of Guelph

Dr. C.K. Minns
Dept. of Fisheries and Oceans
Bayfield Institute (GLLFAS)
Burlington, Ontario

Dr. Michael Moss
University of Guelph

Gary Mouland
Canadian Parks Service

Jo-Anne Moull
Natural Heritage Stewardship Program
Dept. of Land Resource Science
University of Guelph

Dr. Ted Munn
Institute for Environmental Studies
University of Toronto

J. Nelson
Corporation of the City of Kitchener

G. Nixon
Corporation of the City of Kitchener

Dr. Tom Nudds
Zoology Department
University of Guelph
Guelph, Ontario

Tim O'Brien
Corporation of the City of Kitchener

Peter Ottesen
National Parks Systems Branch
Canadian Parks Service
Ottawa, Ontario

Paule Ouellet
Department of Geography
University of Waterloo

Dr. Ajith Perera
Ontario Ministry of Natural Resources
Forest Management Branch
Sault Ste. Marie, Ontario

Joseph Potvin
Consultant
Ottawa, Ontario

K. Pounder
Niagara Escarpment Commission
Georgetown, Ontario

Kathryn Pounder
Niagara Escarpment Commission
Georgetown, Ontario

J.P. Prevelt
Ontario Ministry of Natural Resources
London, Ontario

Dr. Peter Quinby
Toronto, Ontario

Dr. George Quinby
Toronto, Ontario

Debbie Ramsay
Niagara Escarpment Commission
Georgetown, Ontario

Dr. Henry Regier
Institute for Environmental Studies
University of Toronto
Toronto, Ontario

Dr. Trefor Reynoldson
Environment Canada
Burlington, Ontario

Gail Rhynard
Campden Landscaping
Toronto, Ontario

Clay Rubec
North American Wetlands Conservation Council (Canada)
Ottawa, Ontario

Vicki Sahanatien
Canadian Parks Service
Ottawa, Ontario

Todd Salter
Town of Caledon

Andre Savoie
Chief, Resource Management
National Parks Directorate
Canadian Parks Service
Hull, Quebec

Nancy Schackle
School for Resource and Environmental Studies
Dalhousie University
Halifax, Nova Scotia

Eric Schneider
Chief Scientist
Department of Commerce
National Oceanic and Atmospheric Admin.
Washington, D.C. U.S.A.

Elizabeth Seale
Canadian Parks Service

Rafal Serafin
Heritage Resources Centre
University of Waterloo

Mirek J. Sharp
Geomatics International
Guelph, Ontario

Ish Singhal
Agriculture Canada
Sustainable Agriculture
Ottawa, Ontario

B. Sleeth
Corporation of the City of
Kitchener

Peter G. Sly
Rawson Academy of Aquatic
Science
Ottawa, Ontario

Dr. Robert Steedman
Ministry of Natural Resources
Thunder Bay, Ontario

Bill Stephenson
Canadian Parks Service
Cornwall, Ontario

Carolyn Stewart
Niagara Escarpment Commission
Georgetown, Ontario

Lee Swanson
Department of Geography
University of Waterloo

Graeme Swanwick
Ministry of Natural Resources
North Bay, Ontario

Kimberly Taylor
Ministry of Natural Resources
Carleton Place, Ontario

Marion Taylor
Federation of Ontario Naturalists
Don Mills, Ontario

Brent Tegler
University of Guelph

John Theberge
Faculty of Environmental Studies
University of Waterloo

B. Trushinski
City of Waterloo

David Turner
Town of Caledon
Caledon, Ontario

Dr. Tom Whillans
Environment and Resource Studies
Trent University

Carol Whitfield
Canadian Parks Service
Senior Management Advisor's
Office
Environment Canada
Ottawa, Ontario

Peter Wigham
The Metropolitan Toronto &
Region C.A.
Downsview, Ontario

Stephen Woodley
Canadian Parks Service
Ottawa, Ontario

Chi-Ling Yeung
Department of Geography
University of Waterloo